Non-Thermal Processing Technologies for the Dairy Industry

Non-Thermal Processing Technologies for the Dairy Industry

Edited by
M. Selvamuthukumaran
Sajid Maqsood

CRC Press is an imprint of the
Taylor & Francis Group, an **informa** business

First edition published 2022
by CRC Press
6000 Broken Sound Parkway NW, Suite 300, Boca Raton, FL 33487-2742

and by CRC Press
2 Park Square, Milton Park, Abingdon, Oxon, OX14 4RN

© 2022 selection and editorial matter, M. Selvamuthukumaran and Sajid Maqsood; individual chapters, the contributors

CRC Press is an imprint of Taylor & Francis Group, LLC

Reasonable efforts have been made to publish reliable data and information, but the author and publisher cannot assume responsibility for the validity of all materials or the consequences of their use. The authors and publishers have attempted to trace the copyright holders of all material reproduced in this publication and apologize to copyright holders if permission to publish in this form has not been obtained. If any copyright material has not been acknowledged please write and let us know so we may rectify in any future reprint.

Except as permitted under U.S. Copyright Law, no part of this book may be reprinted, reproduced, transmitted, or utilized in any form by any electronic, mechanical, or other means, now known or hereafter invented, including photocopying, microfilming, and recording, or in any information storage or retrieval system, without written permission from the publishers.

For permission to photocopy or use material electronically from this work, access www.copyright.com or contact the Copyright Clearance Center, Inc. (CCC), 222 Rosewood Drive, Danvers, MA 01923, 978-750-8400. For works that are not available on CCC please contact mpkbookspermissions@tandf.co.uk

Trademark notice: Product or corporate names may be trademarks or registered trademarks and are used only for identification and explanation without intent to infringe.

ISBN: 978-0-367-67517-2 (hbk)
ISBN: 978-1-032-11724-9 (pbk)
ISBN: 978-1-003-13871-6 (ebk)

DOI: 10.1201/9781003138716

Typeset in Kepler Std
by Deanta Global Publishing Services, Chennai, India

Dedication

I profoundly thank

God

my family

my friends

and

everyone

who has inspired, supported and wholeheartedly
encouraged me to complete this book

M. Selvamuthukumaran

Contents

Preface ix
Editors xi
Contributors xiii

1 Novel Approach to Non-Thermal Applications in the Dairy
Processing Industry 1
M. Selvamuthukumaran and Sajid Maqsood

2 Pulsed Electric Field Processing of Milk 11
Nidhi Bansal, Farzan Zare, Negareh Ghasemi and Jie Zhang

3 High Hydrostatic Pressure Processing for Dairy Products 35
M. Selvamuthukumaran, Nilesh Nirmal and Sajid Maqsood

4 Cold Plasma: An Emerging Technology in Milk and
Dairy Product Processing 43
Dharini Manoharan and Mahendran Radhakrishnan

5 UV Pasteurization Technology Approaches for Market Milk Processing 67
*Nazia Nissar, Sadaf Rafiq, Rabia Latif, M. Yaseen Sofi, Taibah Bashir
and Sheikh Mansoor*

6 Application of Ultrasound for Dairy Product Processing 81
Maryam Enteshari, Collette Nyuydze and Sergio I. Martinez-Monteagudo

7 Supercritical CO_2 Extraction Process 93
*Kaavya Rathnakumar, Ahmed Hammam, Juan Camilo Osorio,
and Sergio I. Martinez-Monteagudo*

8 Application of Radiation-Based Processing of Dairy Products 105
M. Selvamuthukumaran and Sajid Maqsood

9 Bio-Preservation of Dairy Products: A Non-Thermal Processing and
  Preservation Approach for Shelf-Life Extension of Dairy Products   111
  Nilesh Prakash Nirmal and Chalat Santivarangkna

10 Treatment of Dairy Industry Wastewater with Non-Thermal
   Technologies   127
   Maryam Enteshari and Sergio I. Martinez-Monteagudo

11 Surface Pasteurization and Disinfection of Dairy Processing
   Equipment Using Cold Plasma Techniques   143
   Kamalapreetha Baskaran and Mahendran Radhakrishnan

12 Safety, Regulatory Aspects and Environmental Impacts of
   Using Non-Thermal Processing Techniques for Dairy Industries   157
   Khalid A. Alsaleem, Ahmed R. A. Hammam and Nancy Awasti

   Index   173

Preface

Dairy industries usually adopt conventional methods of processing various milk-based food products, which may actually destroy several nutrients and minimize organoleptic qualities. An alternative approach to this traditional processing technique leads to the use of a non-conventional method, i.e. non-thermal processing techniques, which cannot only enhance consumer acceptability but also enhance the nutritional profile of the various processed products. There are some emerging non-thermal processing techniques such as pulsed light, cold plasma, high-pressure processing, ultrasound, UV pasteurization and ozone treatments that can be successfully employed in dairy processing industries to enhance product acceptability, safety and quality aspects.

This book describes several emerging non-thermal processing techniques, which can be specially employed for dairy processing industries. The book will narrate the benefits of using pulsed light, cold plasma, high pressure and ultrasound during the processing of various dairy products and portray the scope and significant importance of adopting UV pasteurization in processing market milk along with its safety and environmental impacts. The book will address techniques used for the extraction of functional food components from various dairy products by using supercritical CO_2 extraction technology. It explains the application of ozone and cold plasma technology for treating dairy processing wastewater with its efficient recycling aspects. It describes the importance of using bio-preservatives in the shelf-life extension of several dairy food products. This book will be a ready guide for dairy technologists for sterilizing the various dairy process equipment by using microwave cold plasma techniques. It will also address various safety and environmental issues regarding implementation of the various non-thermal processing approaches in dairy processing industries and enlighten dairy technologists, dairy scientists, dairy engineers, professors, research scholars and students with respect to emerging techniques in non-thermal applications for the dairy sector.

I would like to express my sincere thanks to all the contributors; without their continuous support, this book would not have seen daylight. We would like also to express our gratitude toward Mr. Steve Zollo and all the other CRC Press people, who have made every effort to make this book a great standard publication at a global level.

M. Selvamuthukumaran

Editors

M. Selvamuthukumaran is presently Associate Professor and Head of the Department of Food Technology, Hindustan Institute of Technology and Science, Chennai, India. He was a visiting Professor at Haramaya University, School of Food Science and Postharvest Technology, Institute of Technology, Dire Dawa, Ethiopia. He received his PhD in Food Science from the Defence Food Research Laboratory affiliated with the University of Mysore, India. His core area of research is the processing of underutilized fruits for the development of antioxidant-rich functional food products. He has transferred several technologies to Indian firms as an outcome of his research work. He has received several awards and citations for his research and published several international papers and book chapters in the area of antioxidants and functional foods. He has guided several national and international postgraduate students in the area of food science and technology.

Sajid Maqsood is Associate Professor and Assistant Dean for Research and Graduate Studies at the College of Food and Agriculture, United Arab Emirates University, UAE. He teaches advance courses in food science and food chemistry.

His major interests are in food protein and lipid chemistry, bioactive peptides, functional foods and nutraceuticals, plant-based proteins, novel non-thermal technologies and food byproduct valorization. Currently, he is leading an active research program at UAEU that focuses on bioactive molecules in camel milk and dates, investigating their functional, structural and nutraceutical properties, and their encapsulation at micro and nano levels.

He has published his research in leading reviewed journals and presented his research results at both national and international conferences. He has published more than 70 articles and edited three books. He has received multiple prestigious awards that recognized his research activities and leadership skills. He has taught, supervised and trained several graduate and undergraduate students as well as international researcher scholars in the areas of food protein chemistry, bioactive peptides and proteomics.

Contributors

Khalid A. Alsaleem
Dairy and Food Science Department
South Dakota State University
Brookings, South Dakota, USA

and

Department of Food Science and Human Nutrition
Qassim University
Buraydah, Saudi Arabia

Nancy Awasti
Lactalis American Group
Nampa, Idaho, USA

Nidhi Bansal
School of Agriculture and Food Sciences
The University of Queensland
St. Lucia, Australia

Taibah Bashir
Division of Plant Pathology
Sher-e-Kashmir University of Agricultural Sciences and Technology
Srinagar, India

Kamalapreetha Baskaran
Centre of Excellence in Non-Thermal Processing
Indian Institute of Food Processing Technology
Thanjavur, India

Maryam Enteshari
Dairy and Food Science Department
South Dakota State University
Brooking, South Dakota, USA

Negareh Ghasemi
School of Information Technology and Electrical Engineering
The University of Queensland
St. Lucia, Australia

Ahmed Hammam
Dairy and Food Science Department
South Dakota State University
Brookings, South Dakota, USA

and

Dairy Science Department
Assiut University
Assiut, Egypt

Rabia Latif
Division of Plant Pathology
Sher-e-Kashmir University of Agricultural Sciences and Technology
Srinagar, India

Dharini Manoharan
Centre of Excellence in Non-Thermal Processing
Indian Institute of Food Processing Technology
Thanjavur, India

Sheikh Mansoor
Division of Biochemistry
Sher-e-Kashmir University of Agricultural Sciences and Technology
Jammu, India

Sajid Maqsood
College of Food and Agriculture
United Arab Emirates University
Dubai, United Arab Emirates

Sergio I. Martinez-Monteagudo
Family and Consumer Sciences
and
Department of Chemical and Materials Engineering
New Mexico State University
Las Cruces, New Mexico, USA

Nilesh Prakash Nirmal
Institute of Nutrition
Mahidol University
Bangkok, Thailand

Nazia Nissar
Division of Food Technology
Sher-e-Kashmir University of Agricultural Sciences and Technology
Srinagar, India

Juan Camilo Osorio
BIOLI Research Group
Faculty of Pharmaceutical and Food Science
Universidad de Antioquia
Medellin, Colombia

and

Corporacion Universitaria Minuto de Dios - UNIMINUTO Sede Presencial Research Direction
Bogota DC, Colombia

Mahendran Radhakrishnan
Centre of Excellence in Non-Thermal Processing
Indian Institute of Food Processing Technology
Thanjavur, India

Sadaf Rafiq
Division of Floriculture and Landscape Architecture
Sher-e-Kashmir University of Agricultural Sciences and Technology
Srinagar, India

Kaavya Rathnakumar
Dairy and Food Science Department
South Dakota State University
Brookings, South Dakota, USA

Chalat Santivarangkna
Institute of Nutrition
Mahidol University
Bangkok, Thailand

M. Selvamuthukumaran
Department of Food Technology
Hindustan Institute of Technology and Science
Chennai, India

Yaseen Sofi
Division of Plant Pathology
Sher-e-Kashmir University of Agricultural Sciences and Technology
Srinagar, India

Farzan Zare
School of Information Technology and Electrical Engineering
The University of Queensland
St. Lucia, Australia

Jie Zhang
School of Agriculture and Food Sciences
The University of Queensland
St. Lucia, Australia

Chapter 1

Novel Approach to Non-Thermal Applications in the Dairy Processing Industry

M. Selvamuthukumaran and Sajid Maqsood

CONTENTS

1.1	Introduction	1
1.2	Ultraviolet Light	1
1.3	Ultrasonication	3
1.4	Cold Plasma	4
1.5	Low-Temperature High-Pressure Processing	4
1.6	Pulsed Light	5
1.7	Conclusions	6
References		6

1.1 INTRODUCTION

Non-thermal processing is an effective way to kill microbes and enhance the stability of dairy products. This processing has added advantages over conventional methods of processing dairy foods. This method can destroy pathogenic microbes; thereby it retains the nutritional as well as organoleptic characteristics of the dairy food products. The various non-thermal processing techniques that can be used in dairy processing industries are listed in Figure 1.1 and described in the following sections.

1.2 ULTRAVIOLET LIGHT

Ultraviolet light (UV-C) treatment is a prominent non-thermal processing technique employed to kill microbes in fluid milk without subjecting the milk to heat (Bintsis et al., 2000; Matak et al., 2005; Rossitto et al., 2012; Cappozzo et al., 2015; Crook et al., 2015; Krishnamurthy et al., 2007; Christen et al., 2013; Bandla et al., 2012; Gunter-Ward et al., 2018; Alberini et al., 2015; Koutchma et al., 2019). This technique has a lot of advantages over conventional heat pasteurization, which can reduce flavor and cause nutrient loss, and this process seems to be highly energy efficient.

DOI: 10.1201/9781003138716-1

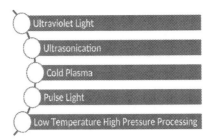

Figure 1.1 Non-thermal processing applications in dairy processing industries.

The drawback of this technique is that treatment results in a lesser reduction in microbes because of its penetration into opaque liquids (Ansari et al., 2019). This problem can be overcome by using turbulent flow reactors, which allow the fluid milk to be exposed to UV light uniformly.

The electromagnetic spectrum of the UV radiation ranges from 100–400 nm, which can be classified into three ranges based on biological effects as well as photochemical properties, i.e. UV-A, B and C (315–400 nm, 280–315 nm, 200–280 nm) (Bintsis et al., 2000; Martysiak-Zurowska et al., 2017). The photochemical changes that occur in nucleic acids and proteins within the cell membrane when fluid milk is subjected to UV-C treatment lead to the inactivation of microbes present in the sample. The cells will die due to the interaction of photons with cystine and thymine nucleoside bases, which leads to the formation of cross-linked photo products like cyclobutyl pyrimidine dimers, which can stop DNA translation, transcription and replication processes, which results in the loss of microbes' cell function and finally leads to microbial cell death (Martysiak-Zurowska et al., 2017; Gayán et al., 2013).

In food industries, generally UV-C light in the wavelength ranged from 250 to 260 nm is being used for the inactivation of microbes like bacteria, viruses, mold, yeast and bacterial spores (Crook et al., 2015).

The use of conventional heat pasteurization techniques for fluid milk has several limitations like higher energy cost, protein denaturation on account of high heat, deterioration of the milk's technological properties, nutrient loss and flavor degradation. Therefore, owing to this effect, implementing non-thermal processing techniques is necessary to retain the biological and functional properties of fluid milk. On this occasion, UV-C light technology plays a predominant role in retaining the above properties without causing any serious technological effects in the samples.

Papademas et al. (2021) studied UV-C light's technological effect on various food-borne microbes, which are artificially inoculated in donkey milk. The various food-borne microbes inoculated in the donkey milk in their studies were *S. aureus*, *B. cereus*, *L. inoccua*, *E. coli*, *Salmonella enteritidis* and *Cronobacter sakazakii*. The inoculated milk was treated with a UV-C dose up to the maximum extent of 1,300 J/L. Except for *L. innocua*, all other food-borne pathogens were destroyed by the treatment exposure to UV-C light at 200–600 J/L. They concluded that UV-C light is a promising non-thermal technology that can be commercially exploited to safeguard raw milk against various food-borne or food spoilage microbes.

The count of *L. monocytogenes* was not reduced and exhibited greater UV-resistant power, which may be ascribed to possession of a thick peptidoglycan cell wall that will prohibit UV photon penetration within bacterial cells; also, *L. monocytogenes* can naturally cope with DNA damage with better DNA repair mechanisms as compared to other food-borne microbes like *E. coli*

(Baysal, 2018; Cheigh et al., 2012; Beauchamp and Lacroix, 2012). In bovine milk, a very high dose of UV at 2,000 J/L was required to reduce *L. monocytogenes* by a 5 log reduction. With respect to goat milk, a 5 log reduction of *L. monocytogenes* was achieved when the goat milk was subjected to a UV dose of 15.8 mJ/cm^2 (Matak et al., 2005; Crook et al., 2015; Lu et al., 2011). The variation in UV dose level for the log reduction of microbes is attributed to the presence of various compositions of fat and total solids in different milk samples like goat, bovine and donkey.

1.3 ULTRASONICATION

Ultrasonication is simply sound waves that can exhibit high frequencies of more than 20 kHz. It has got its wide applications in food processing industries; the application can vary based on its usage, which may be either low- or high-energy ultrasounds. The frequency of the low ultrasound will be 2–3 MHz with intensities of less than 1 Wcm^{-2}. It can be used for detecting foreign particles in both raw material and processed products, and it can also be used for characterizing various food samples (Rezek Jambrek et al., 2010). For the dairy industry, ultrasound with high-energy frequency is preferred, i.e. 18–100 kHz with a sound intensity of greater than 1 Wcm^{-2}.

Several researchers reported that this kind of high-energy ultrasound can be potentially applied in dairy processing industries for the homogenization of milk, inactivation of bacterial enzymes and extraction of chymosin and β-galactosidase (Rezek Jambrak et al., 2009; Jelicic et al., 2010; Bosiljkov et al., 2011). The mechanism involved in ultrasound application lies in the formation of cavitation, i.e. the mechanical nature that leads to the formation and implosion of bubbles in liquid (Brncic et al., 2010). The bactericidal effect of ultrasound can be achieved as a result of implosion, which results in high temperatures, i.e. 5,500° C as well as higher pressures of 50 MPa (Villammiel and de Jong, 2000; Piyasena et al., 2003).

The use of ultrasound has a lethal effect on microbes and is widely used for preserving foods (Entezari et al., 2004; Piyasena et al., 2003; Zenker, 2004). The cavitations are of four kinds, which can be classified based on generation mode, viz. optic, particle, acoustic and hydrodynamic. In food applications, only hydrodynamic and acoustic cavitation were used (Gogate and Kabadi, 2009), as they bring physic-chemical changes to the ultrasound-exposed material.

Rezek Jambrak et al. (2011) conducted a study to notice the impact of ultrasound on model system physical properties prepared with whey protein concentrates (WPC) or whey protein isolates (WPI) after incorporating sucrose and milk powder and also without its addition. The samples after sucrose and milk powder incorporation were subjected to high-power ultrasound treatment of 30 kHz frequency for a period of 5–10 min. After treatment, the samples were analyzed for their solubility, foaming and emulsifying properties, thermophysical and rheological properties. Their study shows that the microstreaming and cavitation effects lead to the occurrence of protein denaturation and thereby exhibit greater influence on all the observed properties. It also further reduced the solubility of protein for whey protein concentrate and whey protein isolate samples when compared with untreated samples. The ultrasound-treated samples exhibited significant raise in foam volume after incorporating either sucrose or milk powder. The samples developed with whey protein concentrates and isolates reduced the emulsion stability indices as well as emulsion activity. The treatment of ultrasound for rheological parameters indicates that it doesn't change the flow behavior indices resulting in noticeable changes in consistency coefficients (k). The treatment had further reduced the starting melting as well as freezing temperatures for sucrose and milk powder incorporated samples.

1.4 COLD PLASMA

Cold plasma is a potent novel technique employed in dairy food processing. All kinds of foods can be successfully preserved by using this technique (Deng et al., 2007; Hosseini et al., 2013; Joshi and Gearson, 2011). Plasma is a gas in ionized form with a half-neutral state containing electrons, ions and chargeless particles like molecules, atoms, ultraviolet photons and radicals. The photons and electrons are usually lighter components when compared to other particles. It's a state of matter by which the heavier components, either non-ionized or ionized particles, can enhance the energy content of gas (Bouhdid et al., 2010). The application of plasma leads to achieving or confirming food safety during food processing (Kim et al., 2011). The mechanism that achieves this process is the occurrence of cell injury and death through oxidation of membranes and electrostatic breakdown (Joshi and Gearson, 2011).

In electrostatic breakdown, the electric charge accumulates in the membrane surface, so that the total electric force overcomes the tensile strength, which keeps the structure of the Gram-negative bacteria membrane, and hence the membrane destruction occurs. The tensile strength of the membrane relates to the murein or peptidoglycan layer, which is thicker in Gram-positive bacteria (15–18 nm) compared to Gram-negative (2 nm). In the second mechanism, the cell membrane damage and the cell contents leak out due to the membrane oxidation and the high energetic ions, radicals, and reactive substances produced during the plasma process. Reactive radicals produced directly in the plasma method can penetrate the cell surface and also the reactive oxygen species (ROS) and finally induce damaging effects on cells by reacting with different macromolecules (Joshi and Gearson, 2011). Using cold plasma for microbial decontamination without the thermal side effects can potentially replace the traditional methods (Saba et al., 2013).

1.5 LOW-TEMPERATURE HIGH-PRESSURE PROCESSING

Milk and its products fall under the category of highly perishable commodities and they will get degraded quickly due to the presence of food-borne microbes, improper pasteurization or sterilization techniques, which leads to possible consumer safety issues (Oliver et al., 2005). Therefore, in order to solve such issues, the industries will perform thermal-based processing techniques, which can enhance the longevity of milk products, thereby destroying the spoilage-causing pathogenic microbes to a greater extent. But with respect to its quality aspects, this process brings changes like protein denaturation, destruction of vitamins and organoleptic changes for sugar-rich milk products. Therefore, as a replacement, it is necessary to choose a non-thermal technique, which can enhance the stability of milk products, thereby destroying pathogens and preserving nutrients. Therefore, it is essential to adopt any new non-thermal technologies for the benefit of consumers as well as food industry professionals (Rahaman et al., 2016; Mauron, 1990; Esteghlal et al., 2019).

One option for a non-thermal processing technique is low-temperature high-pressure processing, which can be a replacement for conventional heat pasteurization and sterilization techniques. Low-temperature high pressure usually adopts a higher pressure application with a lesser temperature, even subzero, i.e. a frozen temperature condition.

Kim et al. (2008) reported that only a few reports were available of successful milk preservation by using the high-pressure processing technique. Such studies justified that pathogen-free safe milk can be produced with fresh attributes. It was observed that the high-pressure processing techniques alone won't destroy spore-forming microbes, which can lead to milk spoilage

(Stratakos et al., 2019; Lee and Kaletunç, 2010). The sterilization action can be achieved by adopting combined different actions like pressure enhancement, the extension of the holding period, the application of heat and preservation with biopreservatives, i.e. nisin, which gave prominent results in stability extension (Stratakos et al., 2019; Lee et al., 2020; Aouadhi et al., 2013; Buzrul, 2017). It was reported that pathogenic microbes can be destroyed at greater than 600 MPa pressure (Balasubramaniam et al., 2008). If we adopt lesser pressure, then there is a necessity to increase the holding period to even higher than 30 min so that pathogens can be inactivated (Buzrul, 2017). This kind of processing can lead to investing more, which will add up maintenance costs, enhance wear and tear, and lead to reduced life of processing equipment with lesser productivity (Torres and Velazquez, 2005). Even though if we use a pressure of 350 MPa for 2 hrs or 400 MPa for 30 min, that won't successfully inactivate microbes in foods, processing time still needs to be reduced as it is important for the food processor (Bermúdez-Aguirre and Barbosa-Cánovas, 2011).

Li et al. (2020) evaluated the low-temperature high-pressure treatment effect, i.e. −25° C and 100 to 400 MPa on frozen milk phase transition behavior. They used specific media to identify the growth of *E. coli*, thereby analyzing the injurious and lethal effects as a result of the application of different pressures to milk. Their results projected that the application of pressure above 300 MPa had induced phase transition from ice I to ice III. The lethal effect was achieved as a result of treatment of 300 MPa at −25° C. It was observed that enhancing the pressure cycle numbers had increased the lethal effects, but it resulted in a slight transformation of injured cells to dead cells. The phase transition had led to the breakdown of cell walls and cell membranes.

1.6 PULSED LIGHT

One of the non-thermal processing techniques to destroy microorganisms is the application of higher-intensity short burst light pulses. This technique is synonymous with various techniques like high-intensity pulsed UV light (HIPL), intense light pulse (ILP), pulsed ultraviolet light (PUV), high-intensity broad-spectrum pulsed light (BSPL) or pulsed white light (PWL) (Palmieri and Cacace, 2005; Krishnamurthy et al., 2010). Microbes like *Listeria monocytogenes* can be successfully killed within a fraction of seconds by applying this technique (Pollock et al., 2017). It will enhance the stability of products to a greater extent (Dunn et al., 1995; Dunn, 1996). This technique can be used to sterilize packaging materials (Buchovec et al., 2010) and also surfaces of the equipment (Woodling and Moraru, 2007; Rajkovic et al., 2010). The only limitation of this technique is because of the food products' opacity and as well as non-uniform surfaces and temperature rises, which can lead to organoleptic quality deterioration.

The US Food and Drug Administration (FDA) has authorized the use of PL technology for processing, handling and also decontamination of food contact surfaces (FDA, 1996). They prescribed the use of a xenon lamp with wavelength (λ) surface-emission ranges from 200 to 1,100 nm, with cumulative treatment not greater than 12 J/cm^2 and pulse width of not more than 2 ms (Palmieri and Cacace, 2005). This technique can be exploited for processing various dairy products in order to kill microbes as shown in Table 1.1.

The light energy is released as a light burst with a shorter duration in a highly concentrated manner (which can even last for a few hundred microseconds, ranging between 1 µs and 0.1 s) (Elmnasser et al., 2007; Abida et al., 2014). The produced light pulses were only short-lived with higher intensity, which accounts for around 20,000 times the sunlight intensity at sea level (Pollock et al., 2017; Dunn et al., 1995; Elmnasser et al., 2007).

TABLE 1.1 APPLICATIONS OF PULSED LIGHT TECHNOLOGY IN DAIRY PRODUCTS

Name of the Dairy Product	Pulse Energy Used (J/cm^2)	Exposure Time (μs)^g	Microbial Count Reduction Log$_{10}$	Reference
Milk	25	114 s	> 2.0 of *Serratia marcescens*	Smith et al. (2002)
Infant milk foods	—	9,500	3 of *L. monocytogenes*	Choi et al. (2010)
Evaporated milk	8.4	—	< 1 of *E. coli*	Miller et al. (2012)
Cheese	53.4	40 s	1.25 of *P. roqueforti* 2.98 of *Listeria monocytogenes*	Can et al. (2014)

Source: modified from Mandal et al., 2020.

1.7 CONCLUSIONS

Therefore, dairy food products can be successfully processed by using several non-thermal processing techniques; thereby the acceptable quality of the processed dairy food products can be produced and supplied to the consumers. Companies can explore such technologies to preserve and protect the nutrients at the same time as enhancing the stability of the processed dairy products to a greater extent.

REFERENCES

Abida, J.; Rayees, B.; Masoodi, F.A. Pulsed light technology: A novel method for food preservation. *Int. Food Res. J.* 2014, 21, 839–848.

Alberini, F.; Simmons, M.J.; Parker, D.; Koutchma, T. Validation of hydrodynamic and microbial inactivation models for UV-C treatment of milk in a swirl-tube 'SurePure Turbulator™'. *J. Food Eng.* 2015, 162, 63–69.

Aouadhi, C.; Simonin, H.; Mejri, S.; Maaroufi, A. The combined effect of nisin, moderate heating and high hydrostatic pressure on the inactivation of Bacillus sporothermodurans spores. *J. Appl. Microbiol.* 2013, 115(1), 147–155.

Ansari, J.A.; Ismail, M.; Farid, M. Investigate the efficacy of UV pretreatment on thermal inactivation of Bacillus subtilis spores in different types of milk. *Innov. Food Sci. Emerg. Technol.* 2019, 52, 387–393.

Balasubramaniam, V.M.; Farkas, D.; Turek, E.J. Preserving foods through high-pressure processing. *Food Technol.* 2008, 62, 32–38.

Bandla, S.; Choudhary, R.; Watson, D.G.; Haddock, J. Impact of UV-C processing of raw cow milk treated in a continuous flow coiled tube ultraviolet reactor. *Agric. Eng. Int. CIGR J.* 2012, 14, 86–93.

Baysal, A.H. Short-wave ultraviolet light inactivation of pathogens in fruit juices. In: Gaurav Rajauria; Brijesh K. Tiwari (Eds.), *Fruit Juices*; Elsevier: Amsterdam, The Netherlands, 2018, 463–510.

Beauchamp, S.; Lacroix, M. Resistance of the genome of Escherichia coli and Listeria monocytogenes to irradiation evaluated by the induction of cyclobutane pyrimidine dimers and 6–4 photoproducts using gamma and UV-C radiations. *Radiat. Phys. Chem.* 2012, 81(8), 1193–1197.

Bermúdez-Aguirre, D.; Barbosa-Cánovas, G.V. An update on high hydrostatic pressure, from the laboratory to industrial applications. *Food Eng. Rev.* 2011, 3(1), 44–61.

Bintsis, T.; Litopoulou-Tzanetaki, E.; Robinson, R.K. Existing and potential applications of ultraviolet light in the food industry–a critical review. *J. Sci. Food Agric.* 2000, 80(6), 637–645.

Bosiljkov, T.; Tripalo, B.; Brnčić, M.; Ježek, D.; Karlović, S.; Jagušt, I. Influence of high intensity ultrasound with different probe diameter on the degree of homogenization (variance) and physical properties of cow milk. *Afr. J. Biotechnol.* 2011, 10(1), 34–41.

Bouhdid, S.; Abrini, J.; Amensour, M.; Zhiri, A.; Espuny, M.J.; Manresa, A. Functional and ultrastructural changes in pseudomonas aeruginosa and staphylococcus aureus cells induced by Cinnamomum verum essential oil. *J. Appl. Microbiol.* 2010, 109(4), 1139–1149.

Brnčić, M.; Karlović, S.; Rimac Brnčić, S.; Penava, A.; Bosiljkov, T.; Ježek, D.; Tripalo, B. Textural properties of infra-red dried apple slices as affected by high power ultrasound pre-treatment. *Afr. J. Biotechnol.* 2010, 9(41), 6907–6915.

Buchovec, I.; Paskeviciute, E.; Luksiene, Z.B. Photosensitization-based inactivation of food pathogen Listeria monocytogenes in vitro and on the surface of packaging material. *J. Photochem. Photobiol. B* 2010, 99(1), 9–14.

Buzrul, S. Evaluation of different dose-response models for high hydrostatic pressure inactivation of microorganisms. *Foods* 2017, 6(9), 79.

Can, F.O.; Demirci, A.; Puri, V.M.; Gourama, H. Decontamination of hard cheeses by pulsed UV-light. *J. Food Prot.* 2014, 77(10), 1723–1731.

Cappozzo, J.C.; Koutchma, T.; Barnes, G. Chemical characterization of milk after treatment with thermal (HTST and UHT) and nonthermal (turbulent flow ultraviolet) processing technologies. *J. Dairy Sci.* 2015, 98(8), 5068–5079.

Cheigh, C.-I.; Park, M.-H.; Chung, M.-S.; Shin, J.-K.; Park, Y.-S. Comparison of intense pulsed light- and ultraviolet (UVC)-induced cell damage in Listeria monocytogenes and Escherichia coli O157:H7. *Food Control* 2012, 25(2), 654–659.

Choi, M.S.; Cheigh, C.I.; Jeong, E.A.; Shin, J.K.; Chung, M.S. Nonthermal sterilization of Listeria monocytogenes in infant foods by intense pulsed-light treatment. *J. Food Eng.* 2010, 97(4), 504–509.

Christen, L.; Lai, C.T.; Hartmann, B.; Hartmann, P.E.; Geddes, D.T. The effect of UV-C pasteurization on bacteriostatic properties and immunological proteins of donor human milk. *PLOS ONE* 2013, 8(12), e85867.

Crook, J.A.; Rossitto, P.V.; Parko, J.; Koutchma, T.; Cullor, J.S. Efficacy of ultraviolet (UV-C) light in a thin-film turbulent flow for the reduction of milkborne pathogens. *Foodborne Pathog. Dis.* 2015, 12(6), 506–513.

Deng, S.; Ruan, R.; Mok, C.K.; Huang, G.; Lin, X.; Chen, P. Inactivation of Escherichia coli on almonds using nonthermal plasma. *J. Food Sci.* 2007, 72(2), M62–M66.

Dunn, J.; Ott, T.; Clark, W. Pulsed-light treatment of food and packaging. *Food Technol.* 1995, 49, 95–98.

Dunn, J. Pulsed light and pulsed electric field for foods and eggs. *Poult. Sci.* 1996, 75(9), 1133–1136.

Elmnasser, N.; Guillou, S.; Leroi, F.; Orange, N.; Bakhrouf, A.; Federighi, M. Pulsed-light system as a novel food decontamination technology: A review. *Can. J. Microbiol.* 2007, 53(7), 813–821.

Entezari, M.H.; Hagh Nazary, S.; Haddad Khodaparast, M.H. The direct effect of ultrasound on the extraction of date syrup and its micro-organisms. *Ultrason. Sonochem.* 2004, 11(6), 379–384.

Esteghlal, S.; Gahruie, H.H.; Niakousari, M.; Barba, F.J.; Bekhit, A.E.-D.; Mallikarjunan, K.; Roohinejad, S. Bridging the knowledge gap for the impact of non-thermal processing on proteins and amino acids. *Foods* 2019, 8(7), 262.

Food and Drug Administration Code of federal. Regulations; 21CFR179.41. FDA: Silver Spring, MD, 1996.

Gayán, E.; Álvarez, I.; Condón, S. Inactivation of bacterial spores by UV-C light. *Innov. Food Sci. Emerg. Technol.* 2013, 19, 140–145.

Gogate, P.R.; Kabadi, A.M. A review of applications of cavitation in biochemical engineering/biotechnology. *Biochem. Eng. J.* 2009, 44(1), 60–72.

Gunter-Ward, D.M.; Patras, A.S.; Bhullar, M.; Kilonzo-Nthenge, A.; Pokharel, B.; Sasges, M. Efficacy of ultraviolet (UV-C) light in reducing foodborne pathogens and model viruses in skim milk. *J. Food Process. Preserv.* 2018, 42(2), e13485.

Hosseini, E.; Asadi, G.H.; Malarreza, L. Maintenance of white meat; 3rd National Conference on Food Security; Savadkooh, Iran, 2013, Sep. 13–15.

Jeličić, I.; Božanić, R.; Tratnik, Lj.; Lisak, K. Possibilities of implementing nonthermal processing methods in the dairy industry. *Mljekarstvo* 2010, 60(2), 113–126.

Jambrak, A.R.; Herceg, Z.; Šubarić, D.; Babić, J.; Brnčić, M.; Brnčić, S.R.; Bosiljkov, T.; Čvek, D.; Tripalo, B.; Gelo, J. Ultrasound effect on physical properties of corn starch. *Carbohydr. Polym.* 2010, 79(1), 91–100.

Joshi, S.; Gearson, S. Nonthermal dielecteric-barrier discharge plasma-induced inactivation involves oxidative DNA damage and membrane lipid peroxidation in Escherichia coil. *Food Sci. Tech.* 2011, 55(3), 1053–1062.

Kim, H.Y.; Kim, S.H.; Choi, M.J.; Min, S.G.; Kwak, H.S. The effect of high pressure–low temperature treatment on physicochemical properties in milk. *J. Dairy Sci.* 2008, 91(11), 4176–4182.

Kim, B.; Yun, H.; Jung, S.; Jung, Y.; Jung, H.; Choe, W.; Jo, C. Effect of atmospheric pressure plasma on inactivation of pathogens inoculated onto bacon using two different gas compositions. *Food Microbiol.* 2011, 28(1), 9–13.

Krishnamurthy, K.; Demirci, A.; Irudayaraj, J. Inactivation of Staphylococcus aureus in milk using flow-through pulsed UV-light treatment system. *J. Food Sci.* 2007, 72(7), M233–M239.

Krishnamurthy, K.; Tewari, J.C.; Irudayaraj, J.; Demirci, A. Microscopic and spectroscopic evaluation of inactivation of Staphylococcus aureus by pulsed UV light and infrared heating. *Food Bioprocess Technol.* 2010, 3(1), 93–104.

Koutchma, T.; Bissonnette, S.; Popovic, V. An update on research, development and implementation of UV and pulsed light technologies for nonthermal preservation of milk and dairy products. *Ref. Module Food Sci.* 2019, 1–21.

Lee, J.; Kaletunç, G. Inactivation of Salmonella enteritidis strains by combination of high hydrostatic pressure and nisin. *Int. J. Food Microbiol.* 2010, 140(1), 49–56.

Lee, Y.-C.; Tsai, Y.-H.; Chen, S.-L.; Kung, H.-F.; Arakawa, O.; Wei, C.-I. Inactivation and damage of histamine-forming bacteria by treatment with high hydrostatic pressure. *Foods* 2020, 9(3), 266.

Li, Y.; Zheng, Z.; Zhu, S.; Ramaswamy, H.S.; Yu, Y. Effect of low-temperature-high-pressure treatment on the reduction of Escherichia coli in milk. *Foods* 2020, 9(12), 1742; doi:10.3390/foods9121742.

Lu, G.; Li, C.; Liu, P. UV inactivation of milk-related microorganisms with a novel electrodeless lamp apparatus. *Eur. Food Res. Technol.* 2011, 233(1), 79–87.

Mandal, R.; Mohammadi, X.; Wiktor, A.; Singh, A.; Singh, A.P. Applications of pulsed light decontamination technology in food processing: An overview. *Appl. Sci.* 2020, 10(10), 3606; doi:10.3390/app10103606.

Martysiak-Żurowska, D.; Puta, M.; Kotarska, J.; Cybula, K.; Malinowska-Pańczyk, E.; Kołodziejska, I. The effect of UV-C irradiation on lipids and selected biologically active compounds in human milk. *Int. Dairy J.* 2017, 66, 42–48.

Matak, K.; Churey, J.; Worobo, R.; Sumner, S.; Hovingh, E.; Hackney, C.; Pierson, M. Efficacy of UV light for the reduction of Listeria monocytogenes in goat's milk. *J. Food Prot.* 2005, 68(10), 2212–2216.

Mauron, J. Influence of processing on protein quality. *J. Nutr. Sci. Vitaminol. (Tokyo)* 1990, 36(Supplement 1), S57–S69.

Miller, B.M.; Sauer, A.; Moraru, C.I. Inactivation of Escherichia coli in milk and concentrated milk using pulsed-light treatment. *J. Dairy Sci.* 2012, 95(10), 5597–5603.

Oliver, S.P.; Jayarao, B.M.; Almeida, R.A. Foodborne pathogens in milk and the dairy farm environment: Food safety and public health implications. *Foodborne Pathog. Dis.* 2005, 2(2), 115–129.

Palmieri, L.; Cacace, D. High intensity pulsed light technology. In emerging technologies for food processing. Sun, D.-W. Ed.; Academic Press: Cambridge, MA; Elsevier Ltd.: Amsterdam, The Netherlands, 2005, 279–306.

Papademas, P.; Mousikos, P.; Aspri, M. Optimization of UV-C processing of donkey milk: An alternative to pasteurization? *Animals (Basel)* 2021, 11(1), 42.

Piyasena, P.; Mohareb, E.; McKellar, R.C. Inactivation of microbes using ultrasound: A review. *Int. J. Food Microbiol.* 2003, 87(3), 207–216.

Pollock, A.M.; Pratap Singh, A.; Ramaswamy, H.S.; Ngadi, M.O. Pulsed light destruction kinetics of L. monocytogenes. *LWT Food Sci. Technol.* 2017, 84, 114–121.

Rahaman, T.; Vasiljevic, T.; Ramchandran, L. Effect of processing on conformational changes of food proteins related to allergenicity. *Trends Food Sci. Technol.* 2016, 49, 24–34.

Rajkovic, A.; Tomasevic, I.; Smigic, N.; Uyttendaele, M.; Radovanovic, R.; Devlieghere, F. Pulsed ultraviolet light as an intervention strategy against Listeria monocytogenes and Escherichia coli O157:H7 on the surface of a meat slicing knife. *J. Food Eng.* 2010, 100(3), 446–451.

Režek-Jambrak, A.; Lelas, V.; Herceg, Z.; Badanjak, M.; Batur, V.; Muža, M. Advantages and disadvantages of high power ultrasound application in the dairy industry. *Mljekarstvo* 2009, 59(4), 267–281.

Režek Jambrak, A.; Lelas, V.; Krešić, G.; Badanjak, M.; Rimac Brnčić, S.; Herceg, Z.; Batur, V.; Grčić, I. Rheological, functional and thermo-physical properties of ultrasound treated whey proteins with addition of sucrose or milk powder. Ultrasound treated whey proteins. *Mljekarstvo* 2011, 61(1), 79–91.

Rossitto, P.; Cullor, J.S.; Crook, J.; Parko, J.; Sechi, P.; Cenci-Goga, B. Effects of UV irradiation in a continuous turbulent flow UV reactor on microbiological and sensory characteristics of cow's milk. *J. Food Prot.* 2012, 75(12), 2197–2207.

Saba, V.; Ramezani, K.H.; Hashemi, H. Bacterial sterilization using dielectric barrier discharge plasma in atmospheric pressure. *MJIRI* 2013, 3, 199–196.

Smith, W.L.; Lagunas-Solar, M.C.; Cullor, J.S. Use of pulsed ultraviolet laser light for the cold pasteurization of bovine milk. *J. Food Prot.* 2002, 65(9), 1480–1482.

Stratakos, A.C.; Inguglia, E.S.; Linton, M.; Tollerton, J.; Murphy, L.; Corcionivoschi, N.; Koidis, A.; Tiwari, B.K. Effect of high pressure processing on the safety, shelf life and quality of raw milk. *Innov. Food Sci. Emerg. Technol.* 2019, 52, 325–333.

Torres, J.A.; Velazquez, G. Commercial opportunities and research challenges in the high pressure processing of foods. *J. Food Eng.* 2005, 67(1–2), 95–112.

Villamiel, M.; de Jong, P. Inactivation of pseudomonas fluorescens and streptococcus thermophilus in Trypticase® Soy Broth and total bacteria in milk by continuous-flow ultrasonic treatment and conventional heating. *J. Food Eng.* 2000, 45(3), 171–179.

Woodling, S.E.; Moraru, C.I. Effect of spectral range in surface inactivation of Listeria innocua using broad-spectrum pulsed light. *J. Food Prot.* 2007, 70(4), 909–916.

Zenker, M. Ultraschall Kombinierte Prozessführung bei der Pasteurisierung und Sterilisierung Flüssiger Lebensmittel, Ph D thesis; Technischen Universität: Berlin; 2004, 149.

Chapter 2

Pulsed Electric Field Processing of Milk

Nidhi Bansal, Farzan Zare, Negareh Ghasemi and Jie Zhang

CONTENTS

2.1	Introduction	12
	2.1.1 Milk	12
	2.1.2 Milk Processing and Need for Alternative Technologies	12
	2.1.3 Alternative Milk Processing Technologies	13
	2.1.3.1 High-Pressure Processing	13
	2.1.3.2 Ultrasonication	14
2.2	Pulsed Electric Field	14
	2.2.1 Introduction	14
	2.2.2 Principle	15
	2.2.3 Mechanisms of Microbial Inactivation	16
	2.2.4 Equipment	18
	2.2.4.1 Pulse Generators	18
	2.2.4.2 Treatment Chamber	20
	2.2.5 PEF in Combination with Other Technologies	21
2.3	Application of PEF to Milk Processing	22
	2.3.1 Composition and Conductivity of Milk Relevant to PEF Processing	22
	2.3.2 PEF Processing of Milk	22
	2.3.2.1 Microbial Inactivation in Milk by PEF	22
	2.3.2.2 Effect of PEF Treatment on Milk Enzymes	24
	2.3.2.3 Effect of PEF Treatment on Vitamins and Volatile Compounds in Milk	25
	2.3.2.4 Effect of PEF Treatment on Milk Lipids, Structure of Milk Fat Globules (MFGs) and Milk Fat Globule Membranes (MFGMs)	26
	2.3.2.5 Effect of PEF Treatment on Milk Proteins	28
2.4	Conclusions	28
References		29

DOI: 10.1201/9781003138716-2

2.1 INTRODUCTION
2.1.1 Milk

Milk is a complex biological fluid that provides nourishment for young mammals from beginning of the life and is considered nature's complete food. It is a complex mixture composed of major components such as water, lipids, proteins, carbohydrates and minerals. In addition, minor components such as vitamins, leucocytes, hormones, organic acids, alcohols, nucleotides, enzymes and non-protein nitrogenous compounds are also found in milk in low concentrations (Walstra et al., 1999; Wong, 2012). Milk and milk products are regarded highly for their nutritional value and play an indispensable role in meeting the nutritional requirements of neonates to adults to the elderly, particularly in Western countries. World milk production was ~852 million tons in 2019 (OECD/FAO, 2020). Cow milk is the most commonly produced milk, accounting for 81% of world milk production. Buffalo (15%), goat, sheep and camel (4% together) milk make up the rest of world milk production. While cow milk is common everywhere, buffalo and camel milk are region specific. Buffalo milk is mainly produced in India while camel milk is mainly produced in the Arabian deserts. India is the largest milk producer in the world, while New Zealand, the European Union and the United States are the three major dairy exporters. China is the world's largest importer of dairy products. Dairy consumption is on the rise in Eastern cultures too, with an increasing per capita consumption rate. Milk production increased by 3.6% in 2019 in China (OECD/FAO, 2020). World per capita consumption of fresh dairy products is projected to increase by 1% per annum over the next decade (OECD/FAO, 2020).

2.1.2 Milk Processing and Need for Alternative Technologies

There is increasing consumer demand for fresh and nutritious food products that are microbiologically safe, with a long shelf-life. Raw milk consumption seems to be increasing in many countries as raw milk has many nutritional components which are believed to be destroyed upon high-temperature processing as well as some organoleptic changes occurring during the heating of milk (Claeys et al., 2013). Although raw milk has a high nutritional value, it can cause serious illness and infectious diseases if it is contaminated with harmful microorganisms. Currently, many treatments are used for the production of packaged milk depending on the final product requirements such as pre-heating, homogenization, separation, heating, cooling and packaging. However, the shelf-life of pasteurized milk and many other processed, non-sterile dairy products is generally limited to one to three weeks. Some combined thermal and physical treatments such as extended shelf-life (ESL) processing can extend shelf-life up to 30 days under refrigeration temperatures while ultra-high temperature (UHT) processed products commonly achieve a shelf life of three months to one year without refrigeration. Antimicrobial agents such as sorbic acid and nisin have been employed in a small number of dairy products to inhibit the growth of undesirable microorganisms but these are not permitted in products such as milk, cream and yogurt. Although the thermal processes have been successfully applied to inactivate pathogenic microorganisms and spores and produce biologically safe milk, they negatively affect the physicochemical properties and nutritional values of milk (Alirezalu et al., 2020). Compared to thermal processes, non-thermal processes have been applied for the efficient decontamination of milk and shown great potential in retaining the essential properties of the milk including flavor, color, density and valuable nutrient compounds as well as the bioactivity of the milk (Alirezalu et al., 2020).

There remains an unmet need for a more effective non-thermal processing aid for bacterial reduction in milk and milk products. In the food industry, any such development would need to be both low-cost and compliant with several strict legal standards and regulations governing product safety and quality.

2.1.3 Alternative Milk Processing Technologies

2.1.3.1 High-Pressure Processing

High-pressure processing (HPP), a non-thermal pasteurization method, is considered a promising alternative to heat pasteurization. This treatment has the potential to inactivate pathogens to provide safe food as well as preserving the quality of the milk. The mechanism of HPP inactivation uses a medium to pass hydrostatic high pressure (usually 400–800 MPa) to food for a short period of time (5–10 min).

HPP has long been used in a wide range of non-food industries, including plastic, metal manufacturing and pharmacy (Bertucco & Vetter, 2001). In 1989, HPP was first applied to milk, which achieved a five- to six-logarithmic reduction in bacteria at 670 MPa after ten minutes. In the 1980s, the first HPP product, fruit jam, was commercialized in Japan. Since then, other foods such as fruit juice, avocados and shellfish have been commercial successes in several countries producing fresh-like, safe foods with long shelf-lives. Nowadays, HPP is being studied for its effect on the functional properties of foods. For example, Fonterra in New Zealand developed a HPP pasteurized fat-free antibody colostrum beverage in 2008 to meet the demands of niche markets. Another example is that Starbucks, the worldwide coffee company, is selling HPP pasteurized fruit juice with premium taste. In 2019, the world's first HPP-treated raw milk was commercialized by a company, Made by Cow, in Australia.

A typical HPP system is made up of a pressure vessel and its closure, a pressure-generating system, a temperature-control system and a material-handling system (Mertens, 1995). Currently, most commercial HPP foods are produced in batches or in a semi-continuous system. After loading the vessel with pre-packed food and closing it firmly, the pressure on the food increases by pumping pressure-transmitting medium into the vessel. Once the target pressure is reached, pumping is stopped and the pressure is held for a set time without applying any further energy. During the processing time, pressure is transmitted uniformly, rapidly and instantaneously to the food and the temperature of the food will increase approximately 3° C per 100 MPa. At the end of the pressure treatment, the vessel is decompressed and the product unloaded.

There are a number of benefits of the high-pressure processing of foods over thermal processing. Although HPP is able to affect the weaker bonds and forces, like electrostatic interactions, hydrogen bridges and van der Walls forces, it does not break covalent bonds. Hence, some properties of food, such as oral sensation, flavor and nutrients, are well preserved. High pressure can be transmitted to food instantaneously, which means it is not time- or mass-dependent. The time required for pressure processing is independent of food geometry and size. HPP eliminates post-processing contamination since food products are pre-packaged before processing, which increases food safety (Mújica-Paz et al., 2011).

However, HPP also has certain limitations, much like any other process. Bacterial spores are resistant to pressure and their elimination requires a combination of processes such as the addition of heat (Smelt, 1998). At present, the high cost also prohibits the widespread use of HPP equipment.

2.1.3.2 Ultrasonication

In recent years, growing attention is being given to the application of ultrasonication (US) in the processing of food such as dairy. This is due to the awareness of its usefulness and advantages in a wide range of applications, as well as the maturity of commercial-scale equipment. At present, the research into US has been focused on improving the processing efficiency and product quality rather than on the inactivation of microorganisms (Ashokkumar et al., 2010).

US is a non-thermal treatment in food application that involves the use of sound waves of high frequency ranging from 18 kHz to 20 MHz. Low-frequency and high-power US, operating at 10–1,000 W/cm^2 and 18–100 KHz, is able to inactivate bacteria as well as change the physical structure of food components (McClements, 1995).

The first report on the effect of ultrasonic inactivation of microbes was in 1929, but the lethal effect on bacteria was limited (Harvey & Loomis, 1929). Advances in US technology over the last decade have triggered interest in bacterial inactivation by US (Pagán et al., 1999). The main principle responsible for this effect is known as cavitation, which involves the growing and imploding of small bubbles in liquid which can consequently generate microjets near the cellular membrane. When sound waves propagate into liquid, they cause alternating high and low pressure. During the pressure changes, these pre-existing bubbles expand, contract and ultimately collapse violently at a critical size. The collapse of bubbles is considered to generate the mechanical force sufficient to break the cell walls of bacteria by rapidly increasing the surrounding temperature and pressure, which can be as high as 5,500° C and 50 MPa respectively (Bermúdez-Aguirre & Barbosa-Cánovas, 2011). But the exact mechanism of US on microbes is still unknown.

It is possible to pasteurize milk at 57° C with an ultrasound frequency of 20 kHz and power intensity of 118 W/cm^2 for 18 minutes (D'amico et al., 2006). Full cream milk requires a longer treatment time than non-cream milk due to the protective action of fat. The effect of US on enzymes in milk is related to the nature of these enzymes, including their heat resistance and molecular structure (Vercet et al., 2001). Three mechanisms contribute to this. The first one is heat produced during cavitation. The second is free radicals created by water sonolysis and the third is mechanical forces generated during microstreaming. It was observed that ultrasound had hardly any effect on lactoperoxidase, alkaline phosphatase and gamma-glutamyltranspeptidase, but inactivation of these three enzymes was identified at 61, 71 and 75.5° C respectively when ultrasound and heat were applied simultaneously (Villamiel & de Jong, 2000). The inactivation extent of lactoperoxidase and alkaline phosphatase increased with increasing power levels, exposure time and temperature (Ertugay et al., 2003).

2.2 PULSED ELECTRIC FIELD

2.2.1 Introduction

Pulsed electric field (PEF) is a powerful high electric field pulse with a short pulse width (in micro- or nanoseconds) and is identified as one of the effective non-thermal treatment processes for decontamination and microorganism inactivation with minimal heat production for a wide range of applications (Ghasemi et al., 2020). The effectiveness of high-intensity pulsed electric fields on living organisms provides novel physical and chemical stress to biological systems, opening a new field known as bioelectrics (Akiyama & Heller, 2017). Such a capability provides tremendous potential in the food industry, where producing high-quality and biologically safe food products is important.

When the high electric fields are applied to the system which includes the biological cells, a potential difference (voltage) will appear across the cell membrane. If the voltage across the cell membrane is greater than the membrane breakdown voltage (> 1 V), electroporation will occur and membrane pores will form. When electroporation occurs, the cell permeability increases. Depending on the intensity of the applied electric field, the poration effect can be reversible or irreversible (Akiyama & Heller, 2017). If a low-intensity electric field with a very short duration is applied, the reversible breakdown of cell membranes will be achieved, which means the poration effect is reversible; however, using a high-intensity electric field with a long duration will result in irreversible electroporation and lead to cell death (Barbosa-Cánovas & Bermúdez-Aguirre, 2010; Schoenbach et al., 2002). Such phenomena play important roles in food processing applications where food decontamination is highly desirable as much as food quality and retaining the nutritional compounds of the food. Several studies showed that PEF treatment could be applied for inactivation of a wide range of microorganisms including *Escherichia coli*, *Salmonella*, *Streptococcus* spp., *Listeria monocytogenes*, *Proteus* spp., *Clostridium* spp., *Lactobacillus* spp., *Pseudomonas* spp. *and Staphylococcus aureus* (Syed et al., 2017). Most of these studies showed a significant reduction in microbial cells after PEF treatment (Sampedro & Rodrigo, 2015; Toepfl et al., 2005). Depending on the characteristics of the microorganisms such as their sizes, shapes, species and strains, the responses of the microbial cells to the PEF treatment will be different. This means the effectiveness of PEF on the inactivation of the microorganism may differ from one microorganism to the other (Syed et al., 2017). For instance, under the same PEF treatment process with similar electrical conditions, the effectiveness of PEF processing on rod-shaped Gram-negative bacteria is different from that on rod-shaped Gram-positive bacteria (Sobrino-López et al., 2006). It has been shown that for the treatment of smaller microbial cells, a larger electric field is required; however, for larger cells, a smaller EF is required. Also, different studies showed that PEF is less effective for the inactivation of Gram-positive bacteria due to their cell wall characteristics (thicker peptidoglycan membrane) (García et al., 2005a). Therefore, for efficient PEF processing, the electrical parameters of the PEF processing should be properly designed.

2.2.2 Principle

In PEF technology, a series of narrow high-voltage pulses are usually generated to deliver a huge amount of energy to the sample under treatment over a short period of time (a few hundredths of a nanosecond/microsecond). As shown in Figure 2.1, although the generated power over a long time is very small (1 watt), the peak power released over a few nanoseconds is huge, around 1 gigawatt.

Considering the electrical characteristics of the solution (load), the amount of delivered energy can be controlled by adjusting the pulse parameters including the amplitude, frequency, duration, rise time and fall time of the pulses as well as the number of applied pulses. Therefore, for an efficient treatment process, the PEF system should be studied and designed at two different levels, the system level and the load level. When high-voltage pulses with a narrow pulse width are applied across two parallel-plate electrodes placed on both side of the solution, an electric field will be generated in the solution which is expressed by Eq. 1:

$$EF = \frac{V}{d} \quad (1)$$

where *EF* represents the generated electric field, *V* is the amplitude of the applied pulses and *d* is the distance between two electrodes.

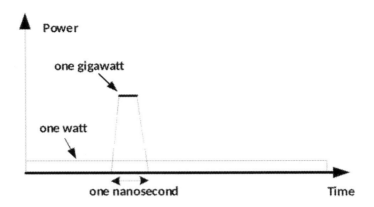

Figure 2.1 Generated and released peak power in pulsed power technology.

As shown in Eq. 1, the intensity of the generated electric field can be adjusted by changing the amplitude of the applied pulses, V, and/or by changing the distance between the two electrodes, d.

2.2.3 Mechanisms of Microbial Inactivation

The phenomena associated with PEF-induced inactivation of microorganisms is due to the increased permeabilization and/or formation of pores in the cellular membrane (Kotnik et al., 2019; Saulis, 2010). Electroporation refers to the formation of aqueous pores on the cell membrane, and electropermeabilization refers to the increased permeability of the membrane which can have many contributing and unknown mechanisms (Kotnik et al., 2019).

Microbial inactivation by pulsed electric fields could be due to several reasons and is dependent on the response of the organism of interest to the physical stimuli within its respective medium (Saulis, 2010). For example, the choice in the duration of the pulse waveform can cause drastically different responses or methods of inactivation. Nanosecond pulsed electric fields in eukaryotic cells triggered a calcium-induced calcium release (positive feedback) of the endoplasmic reticulum, consequently killing the cell (Semenov et al., 2013), whereas microsecond pulsed electric fields caused stronger stresses on the plasma membrane of cells through irreversible electroporation and/or electropermeabilization, consequently killing it (Beebe & Schoenbach, 2005). Pulses that are longer than the membrane charging-time constant strongly affect the plasma membrane, whereas pulses that are shorter than the membrane charging-time constant have a stronger influence on intracellular organelles (Beebe & Schoenbach, 2005). This is dependent on the conductivity and dielectric properties of the liquids (extra- and intracellular media) and membranes (plasma membrane[s] and intracellular membranes) of interest.

Some studies have suggested that bipolar pulses are more advantageous than unipolar pulses due to the movement of charged molecules enhancing the breakdown effect. Moreover, bipolar pulses contribute to less solid deposition on the electrode surface and electrolysis (Kotnik et al., 1997). However, recent studies by Pakhomov et al. mentioned the bipolar cancellation effect which inhibited the permeabilization of the cells (Gianulis et al., 2018; Pakhomov et al., 2018). These studies were conducted using nanosecond pulsed electric fields which have a time constant shorter than the membrane charging time, compared to the previous studies which used pulses in the microsecond range. The bipolar cancellation phenomenon is reported to not get affected by the extracellular conductivity in pulses longer than the membrane charging time (Gianulis et al., 2018). Furthermore, for pulses in the nanosecond range, bipolar cancellation occurrence was higher in lower extracellular conductivity. Pakhomov et al. also reported that bipolar cancellation was not much affected by the pulse width, electric field, spectral content of pulses or the pulse rate. The dominant cause of bipolar cancellation was the ratio of opposite polarity pulses.

Rod-shaped cells were reported to be less PEF susceptible than sphere-shaped cells (Heinz et al., 2001; Qin et al., 1998). However, other studies have reported no significant influence of microorganism shape on PEF inactivation (García et al., 2005b). Microbial inactivation by PEF excitation is dependent on microbial species, strains and serotypes (García et al., 2005b) (Figure 2.2).

For microorganisms such as *Lactobacillus*, which is a genus of Gram-positive, rod-shaped bacteria, it has been shown that nanosecond pulsed electric fields under certain conditions can cause reversibility of their permeabilization (Vaessen et al., 2019). Reversibility of the cell envelope was measured by propidium iodide (PI) and SYTOX green uptake (Vaessen et al., 2019). The delivery of PI is inversely proportional to the extracellular conductivity, where transient electrophoretic transport is reported to be the dominant mechanism and where pore area density and membrane conductance are proportional to the extracellular conductivity (Yu & Lin, 2014).

Some Gram-positive bacteria like *Lactobacillus plantarum* and *Listeria monocytogenes* were shown to be more resistant at a pH of 7.0 as compared to a pH of 4.0 (García et al., 2007), whereas

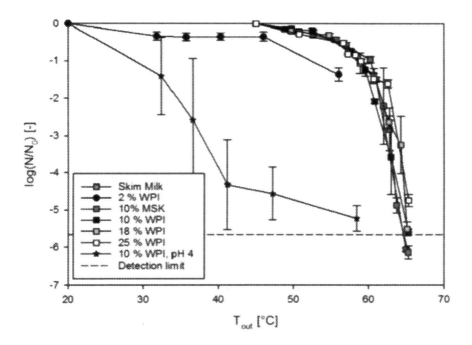

Figure 2.2 Different concentrations of whey protein isolate (WPI) and skim milk concentrate (MSK) at pH 4 and 7 were used to study the PEF inactivation of *Listeria innocua* using a co-linear continuous treatment chamber. Source: directly taken from Figure 2 from Schottroff et al. (2020) with permission.

the Gram-negative bacteria like *E. coli* and *Salmonella senftenberg 775W* were found to be more resistant at a pH of 4.0 (García et al., 2007). Large populations of sublethal injury were noticed in resistant cases using PI staining before and after treatment (García et al., 2005a). However, this study used buffers at fixed conductivities, which did not reflect variations in milk minerals, protein and fat content as noticed between different farms and feeds (García et al., 2005b). Changes in these parameters can strongly influence the PEF-induced inactivation kinetics, as shown in Figure 2.2 (Schottroff et al., 2020).

Fat content was reported to have a protective effect for microbial inactivation when past a threshold, although this was found to be significant for cream, which has a much higher fat content than whole milk (Sampedro & Rodrigo, 2015). Increasing the initial temperature of the milk sample was found to increase the microbial reduction further (Ohshima et al., 2016). After PEF treatment, maintaining the liquid at room temperature, 50° C or 70° C, for up to 20 seconds' duration was also found to contribute to the survival fraction of *E. coli* where longer holding times resulted in higher inactivation (shown in Figure 2.3) (Ohshima et al., 2016). Therefore, for the application of PEF processing in liquid foods, there needs to be an associated "sensitivity analysis" which evaluates the influence of the liquid food variation (i.e. protein concentration, conductivity, fat content, pH) on the PEF treatment process.

2.2.4 Equipment

Each PEF system includes two main parts, the pulse generator and the treatment chamber. Both the pulse generator and the treatment chamber should be properly designed for efficient treatment.

2.2.4.1 Pulse Generators

In pulsed power technology, the pulse's electrical characteristics including pulse amplitude, frequency, pulse width, pulse rise and fall times and the number of the applied pulses play critical roles in the effectiveness of the pulsed power technique in different applications where

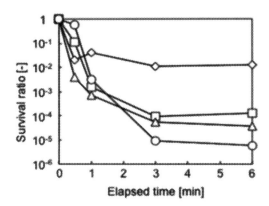

Figure 2.3 Survival ratio of *E. coli* in a continuous PEF treatment chamber (inlet temperature: 50° C, 60 mL/min, 40 kV unipolar exponentially decaying, 100 Hz) with different holding times (time between PEF treatment and cooling to 4° C) at room temperature: 0.5 seconds (diamonds), 5 seconds (squares), 10 seconds (triangles), and 20 seconds (circles). Source: directly taken from Figure 7 from Ohshima et al. (2016) with permission.

inappropriate pulse parameters can degrade the effectiveness of the pulsed power method. Several pulse generators were developed to generate high-voltage pulses with different electrical characteristics (Akiyama & Heller, 2017; Ho & Mittal, 2000). Some of these pulse generators are discussed here. The simplest pulsed power generator is known as a pulse-forming line and consists of two conductors—inner and outer conductors which are isolated from each other (coaxial line)—a DC source, a switch and a resistor connected between the source and the coaxial line. The inner conductor at one end is connected to the DC source through a resistor and at another end is directly connected to the utilized switch and then to the load. Once the switch is closed, the charging voltage appears on the inner inductor. The maximum amplitude of the output voltage appears across a resistive load equivalent to half of the charging voltage. The width of the generated pulses can be adjusted by changing the length of the conductors and the value of the charging voltage (Akiyama & Heller, 2017).

The Blumlein line is another pulse generator that was developed by Blumlein. In this pulse generator, three metal strips run parallel and the middle and bottom strips can be connected through a switch placed parallel to these two strips. The top and bottom strips are connected through the load. A DC voltage source is connected between the middle and bottom strips. The positive terminal of the DC voltage source is connected to the middle strips through a resistor and the negative terminal is directly connected to the bottom strips. Similar to the pulse forming line generators, the width of the generated pulses is directly proportional to the length of the strips and inversely proportional to the charge developed on the middle strip. With a Blumlein configuration, the maximum amplitude of the output pulses is twice that of pulse-forming generators (Akiyama & Heller, 2017; Rebersek et al., 2014).

The capacitor discharge circuits are widely used in pulsed power applications to generate high-voltage exponentially decaying pulses. In these circuits, the utilized capacitor(s) are charged and discharged by a switch in the designed circuits to generate high-voltage pulses. The maximum amplitude of the output pulses is a product of the amplitude of the input DC voltage source and the number of applied capacitors. The Marx generator is one of the most common capacitor discharge circuits which has high voltage and high power ratings but suffers from large energy loss associated with utilizing large resistors and long capacitor charging times. Although the discharge capacitor circuits are simple, generating high-voltage pulses with desired characteristics using the conventional switches (non-solid-state switches) is challenging because of the switches' characteristics that limit the capability of the pulse generators.

With the development of solid-state switches like metal–oxide–silicon field-effect transistors (MOSFETs) and insulated-gate bipolar transistors (IGBTs), several pulse generators like the solid-state Marx generator and power electronic-based pulse generators were developed benefiting from superior characteristics of the solid-state switches, lower power losses and higher efficiency. These solid-state switches could operate at higher voltage ranges (< few kilovolts) without producing additional losses but their main drawbacks were limited switching speed and power rating (Akiyama & Heller, 2017; Rebersek et al., 2014). Among power electronic converters designed for pulsed power systems, flyback, forward, half-bridge and full-bridge converters are widely used (Bochkov et al., 2012). The flyback and forward converters are usually used to generate unipolar pulses; however, half-bridge and full-bridge converters are used to generate bipolar pulses. With the new generation of switching devices, wide-bandgap (WBG) semiconductor devices that can switch very fast and at very high voltage levels (> few kilovolts), high-voltage narrow pulses (nano/picoseconds) with fast rise and fall times can be generated for pulsed power applications.

2.2.4.2 Treatment Chamber

The uniformity in the distribution of treatment intensity is very important in the PEF treatment processes as inhomogeneous distribution of the generated electric fields can contribute to the local temperature rise and degrade the effectiveness of the treatment process (Jäger & Knorr, 2017). To achieve uniform PEF, the treatment chambers should be designed with optimized geometry. Several treatment chambers with different geometric configurations have been developed for static and continuous treatment processes. Most of these treatment chambers consist of at least one high-voltage electrode and one ground electrode which are insulated from each other (Sampedro & Rodrigo, 2015; Toepfl et al., 2005). Several treatment chambers were designed with different electrode configurations including flat parallel electrodes, a co-linear chamber, and a coaxial chamber which are shown in Figure 2.4. Among these configurations, parallel electrodes are normally used to produce the most uniform electric field (Smith et al., 2002). This configuration is mostly used for static or batch treatment processes with small volumes (Sampedro & Rodrigo, 2015; Toepfl et al., 2005). One of the major concerns of using a static treatment chamber with a small volume is the large temperature rise in the liquid sample when large electric fields are generated, which can lead to the degradation of valuable nutrient compounds. This issue can be addressed through proper control and design of pulse parameters such as pulse magnitudes, repetition rate, number of pulses and pulse width.

Compared to static treatment chambers, the design of continuous treatment chambers is more challenging as the fluid dynamic aspects should also be included in the design process to achieve an efficient and effective treatment process. For the continuous treatment processes, several factors can affect the effectiveness of the treatment process including the flow rate, chamber volume and pulse frequency (Sampedro & Rodrigo, 2015). Parallel plate configurations have been also used for continuous treatment processes. In this configuration, the direction of the generated electric field is perpendicular to the direction of the liquid flow and the liquid sample is stimulated by the electric field continuously throughout the treatment zone between

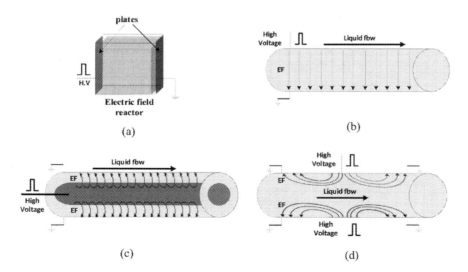

Figure 2.4 Treatment chamber configurations for static treatment: (a) parallel plate and continuous treatment, (b) parallel plate, (c) coaxial and (d) co-linear.

the plates (Ho & Mittal, 2000). For the plate-to-plate configuration, since the electrode surface area is large and the intrinsic electrical resistance is small, a large current will flow through the sample during the treatment process which can contribute to the occurrence of undesired electrochemical phenomena at the electrode-electrolyte surface (Toepfl et al., 2005). In addition to parallel plate configuration, the coaxial and co-linear chambers are used for the continuous treatment process. Compared to the plate-to-plate configuration, in the coaxial chambers, the electric field distribution is inhomogeneous which can degrade the effectiveness of the PEF treatment process. Similar to the plate-to-plate chambers, in co-linear chambers, the directions of the electric fields and liquid flow are perpendicular to each other. However, the electrode surface area is small due to the cylinder shape of the electrode which leads to large intrinsic resistance and small current flows through the liquid. The co-linear chambers produce the least homogeneous electric field distribution but this issue can be addressed by the proper design of insulators and the chambers' geometries (Toepfl et al., 2005).

2.2.5 PEF in Combination with Other Technologies

Pulsed electric fields have been studied in combination with other accompanying techniques such as heating, sonication, sparging and bacteriocins such as nisin for the inactivation of microorganisms and endospores. Destruction of endospores is usually through high-temperature treatment above 120° C. The combination of PEF with heating above 60° C and 30 kV/cm was found to contribute to the inactivation of endospores and has lower energy costs than thermal treatment alone (Siemer et al., 2014).

Combination of high-pressure CO_2 with PEF at low temperature completely inactivated vegetative cells of *B. cereus*, *S. aureus* and *E. coli*, and achieved a 3 log reduction in spores (Spilimbergo et al., 2003). Applying PEF and keeping the liquid in acidic conditions can contribute to enhanced microbial inactivation (García et al., 2005b). PEF treatment combined with sparging (e.g. using CO_2 to de-bubble solution) can be useful as it removes microbubbles which can influence the homogeneity of the electric field and peak amplitude of the applied voltage, and, at high electric field strengths, can cause micro-discharging. Sparging may have an enhanced inactivation effect for aerobic microorganisms when used with PEF.

Matrix constituents, such as lactoferrin, may be able to cause higher inactivation of Gram-positive bacteria as they break down the thick outer layer of peptidoglycan. Human milk contains more lysozyme and lactoferrin than bovine milk. Researchers studied the effect of PEF and PEF-nisin treatment on skimmed bovine milk by visualizing *Listeria innocua* with transmission electron microscopy (TEM) (Calderón-Miranda et al., 1999). Nisin is a food preservative (bacteriocin) produced naturally by the Gram-positive *Lactococcus lactis*. The authors reported an increase in the cell wall's roughness when increasing the electric field intensity. Furthermore, a combination of PEF and nisin was found to induce damage to the cell walls and a slight increase in cell length (Calderón-Miranda et al., 1999).

PEF-induced heating was found to contribute to the inactivation of lactoperoxidase in simulated milk ultrafiltrate (SMUF), where PEF combined with mild heating could lead to increased inactivation as compared to thermal treatment alone (Buckow et al., 2012). PEF at higher starting temperatures has been reported to be more efficient for the inactivation of microorganisms, as compared to lower starting temperatures (Ohshima et al., 2016). For cheese/yogurt application, combining lactic acid bacteria (LAB) that produce bacteriocins with PEF for the inactivation of competing microorganisms may be a useful technique to facilitate inactivation.

2.3 APPLICATION OF PEF TO MILK PROCESSING

2.3.1 Composition and Conductivity of Milk Relevant to PEF Processing

From the various compounds present in milk, such as water, fat, proteins, minerals, vitamins and lactose, minerals are the most dominant contributor to its conductivity (Lawton & Pethig, 1993; Mabrook & Petty, 2003), whereas lactose and casein were found to have a much smaller effect on milk conductivity. It has been reported that the conductivity of milk decreases with an increase in fat content (Lawton & Pethig, 1993; Mabrook & Petty, 2003; Żywica et al., 2012). This is because most of the fat present in milk is surrounded by a non-conductive membrane ranging from 2 to 10 μm in diameter. The relationship between the conductivity and the milk fat content is described by a Lawton and Pethig equation (Eq. 2) (Lawton & Pethig, 1993):

$$G = G_s(1-v)n \qquad (2)$$

where G is the conductivity of milk, G_s is the conductivity of fat-free (skim) milk, v is the volumetric fraction of fat present in milk and n is a constant at 1.7.

Insoluble salts in milk such as calcium phosphate are associated with casein micelles in the colloidal phase where they hold the subunits together (Fox et al., 1998). Storage and acidification of milk have been shown to affect its conductivity. Mabrook and Petty reported that when milk was kept at room temperature for 48 hours, the colloidal calcium phosphate was released in the serum phase of milk and increased the conductivity of milk by 15% (Mabrook & Petty, 2003). Furthermore, they found that acidification of milk to a pH of 5.0 could cause a similar response and increase the conductivity of the milk through gradual solubilization of colloidal calcium salts from the casein micelles.

The interaction of pulsed electric field modalities with casein micelles and fat globules may contribute to the release of the colloidal salts. Therefore, a transient sudden increase in conductivity during pulsed modalities may impact the temperature variations and response of bacterial inactivation through ohmic heating.

Floury et al. and Michalac et al. reported that the pH and electrical conductivity of skim milk did not change after PEF treatment (Floury et al., 2006; Michalac et al., 2003).

2.3.2 PEF Processing of Milk

A thorough search of the relevant literature suggests that the only commercial application of PEF has been for fruit juice by Genesis Juices, Oregon, USA, while only laboratory and pilot-scale studies have been conducted in milk (Lee et al., 2015). There have been several research studies on the evaluation of microbial safety, nutritional quality and enzymatic activity of milk processed by PEF. PEF has been proven to reduce the number of microorganisms existing in milk while having minimal detrimental effects on the quality of the milk (Bendicho et al., 2002a). Considerable research efforts have been devoted to PEF studies using simulated milk ultrafiltrate (SMUF), skim milk and whole milk as the media.

2.3.2.1 Microbial Inactivation in Milk by PEF

In addition to PEF treatment parameters, such as electric field intensity, treatment time, pulse shape, pulse polarization and pulse frequency, the efficiency of microbial inactivation by PEF is also affected by milk properties. Several authors reported that the microbial inactivation by PEF decreased with increased fat content in the medium. It has been speculated that milk fat has a protective effect for bacteria against electric pulses (Grahl & Märkl, 1996). However, Reina et al. (1998) reported that the inactivation rates by PEF treatment were similar among skim milk, 2% fat milk and whole milk. Sobrino-López et al. (2006) also reported that the fat content did not affect the inactivation of the *Staphylococcus aureus* inoculated in milk. The effect of PEF treatment on various bacteria in whole milk is summarized in Table 2.1.

Three studies described a detailed comparison of microbial inactivation under different PEF conditions, in particular a combination of electric field strength, treatment time and pre-heating

TABLE 2.1 THE EFFECT OF PEF TREATMENT ON VARIOUS BACTERIA IN WHOLE MILK

Milk Fat Content	PEF Treatment Condition	Microorganism	$\log_{10}$ Reduction	Reference
4% fat milk	35 kV/cm, 30 µs, 40° C	*Escherichia coli*	1.2–2.5	Walter et al. (2016)
		Pseudomonas	1.4–3.0	
4% fat milk	30 kV/cm, 22 µs, 53° C	Total plate account	0.5	McAuley et al. (2016)
	30 kV/cm, 22 µs, 63° C		0.9	
~4% fat milk	16–26 kV/cm, 34 µs, 55° C	Total plate account	5.2–6.0	Sharma et al. (2014a)
		Escherichia coli	> 6.0	
		Listeria monocytogenes	> 6.0	
3.8% fat milk	26 kV/cm, 67 µs, 40° C	*Pseudomonas aeruginosa*	2.0	Sharma et al. (2014b)
	23 kV/cm, 24 µs, 50° C	*Pseudomonas aeruginosa*	> 6	
	23 kV/cm, 34 µs, 50° C	*Escherichia coli*	2.0	
	28 kV/cm, 34 µs, 50° C	*Staphylococcus aureus*	2.0	
	28 kV/cm, 34 µs, 50° C	*Listeria innocua*	2.0	
Whole milk	30 kV/cm, 50 µs, 62° C	*Listeria innocua*	4.3	Guerrero-Beltrán et al. (2010)
	40 kV/cm, 43.75 µs, 68° C	*Listeria innocua*	5.5	
Raw milk	30 kV/cm, 29 µs, < 60° C	*Lactobacillus rhamnosus*	2.0	Guerrero-Beltrán et al. (2010)
3.4% fat milk	60 kV/cm, 200 µs, 40° C	*Pseudomonas flurescens*	5.0	Shin et al. (2007)
	60 kV/cm, 200 µs, 50° C	*Bacillus stearothermophilus*	3.0	
	60 kV/cm, 200 µs, 40° C	*Escherichia coli*	5.5	
3.0% fat milk	30 kV/cm, 600 µs, < 25° C	*Staphylococcus aureus*	2.1	Sobrino-López et al. (2006)
3.6% fat milk	29 kV/cm, 250 µs, < 45° C	*Listeria innocua*	2.0	Picart et al. (2002)
3.5% fat milk	30 kV/cm, 600 µs, ~25° C	*Listeria innocua*	~3	
	30 kV/cm, 300 µs, ~50° C	*Listeria innocua*	~3	Reina et al. (1998)
	30 kV/cm, 600 µs, ~50° C	*Listeria innocua*	~4	

temperature (Guerrero-Beltrán et al., 2010; Reina et al., 1998; Sharma et al., 2014a). Sharma et al. reported that pre-heating of milk for 24 s over 50° C prior to PEF treatment reduced the number of bacteria to undetectable levels, while pre-heating below 50° C had limited effect on bacterial reduction. This effect is not only because of the thermal-injury effect but also the increased susceptibility of the bacterial membrane to PEF treatment which is synergistic with elevated preheating temperatures (Sharma et al., 2014a; Walter et al., 2016).

Vega-Mercado et al. (1996) has reported the effects of ionic strength and pH of milk on PEF processing. The rate of microbial inactivation was reduced in higher ionic strength solutions. When the pH was reduced from neutral, the rate of inactivation was also increased. On the other hand, Jayaram reported that increased conductivity of the fluid decreased the inactivation effect of bacteria by PEF (Jayaram et al., 1993).

Odriozola-Serrano et al. (2006) did a comparative analysis on how PEF treatment or HTST pasteurization (72° C for 15 s) can impact the shelf-life of whole milk. A continuous flow system with eight colinear chambers placed in series was used for PEF treatment. The electric field strength used was 35.5 kV/cm, while the treatment was 300 or 1,000 μs. It was reported that PEF treatment for 1 ms reduced the initial concentration of the mesophilic aerobic microorganisms from 3.2 log cfu/ml to 2.2 log cfu/ml (1 log reduction), while 0.3 ms treatment time resulted in less than 1 log reduction; thermal pasteurization resulted in 2 log reduction.

2.3.2.2 Effect of PEF Treatment on Milk Enzymes

PEF treatment can affect both the endogenous and exogenous enzymes present in milk. PEF treatment can inactivate, stimulate or cause no changes to the enzyme activities in milk. The effect of PEF on enzyme activity is mainly dependent on the PEF treatment parameters, inherent properties of the enzymes (enzyme structure, type of bonds and interactions, geometry of active site, etc.), properties of the medium (composition, viscosity, ionic strength, pH, conductivity, etc.) as well as temperature of processing (Poojary et al., 2016). Most of the studies on PEF treatment of milk have focussed on microbial inactivation and the information on its effect on milk enzymes is limited. Although the mechanism of effect of PEF on enzyme activity is not fully elucidated, studies suggest that the electrical and thermal phenomena associated with PEF can bring about conformational and structural changes in enzymes, altering their activities. It is suggested that PEF can modify the structure of active sites in enzymes or the overall molecular conformation of the enzymes. PEF can also induce association or dissociation of functional moieties in enzymes, leading to their denaturation. It is suggested that PEF causes limited changes in the primary structure of the enzymes (Ohshima et al., 2007). However, the inactivation of enzymes by PEF is due to the changes in secondary and tertiary structures (Castro et al., 2001a). Plausible mechanisms of enzyme inactivation have been detailed in recent reviews (Poojary et al., 2016; Zhao et al., 2012).

Several researchers have studied the effect of PEF on milk enzymes. Numerous studies focused on the effect of PEF on alkaline phosphatase (Castro et al., 2001b; Castro et al., 2001a; Grahl & Märkl, 1996; Ho et al., 1997; Riener et al., 2009; Shamsi et al., 2008; Sharma et al., 2014b). Other milk enzymes of interest have been plasmin (Qin et al., 1996; Sharma et al., 2014b; Vega-Mercado et al., 1995), lactoperoxidase (Grahl & Märkl, 1996; Riener et al., 2009), xanthine oxidase (Sharma et al., 2014b) and lipase (Grahl & Märkl, 1996; Ho et al., 1997; Riener et al., 2009; Sharma et al., 2014b). Several reviews summarized the studies on the effects of PEF on milk enzymes (Poojary et al., 2016; Van Loey et al., 2001; Zhao et al., 2012). Studies on enzymes have been conducted both in milk and simulated media such as buffers or synthetic milk ultrafiltrate. A range of PEF parameters and processing equipment has been used. Hence, the results from various studies on a particular enzyme are not comparable. Some studies have reported a significant reduction in enzyme activity in milk (Sharma et al., 2014b), while others have elucidated a negligible effect of PEF on milk enzymes (Ho et al., 1997; Riener et al., 2009). The disparity in results of these studies may also suggest that there is potential to tailor the PEF processing conditions to suit specific food applications.

Some studies have reported that combining PEF with mild thermal treatments can enhance the effect of PEF on enzymes (Buckow et al., 2012; Shamsi et al., 2008). A recent review summarizes the studies focussing on the effect of the combination of PEF and mild heating on microorganisms, enzymes' composition and the shelf-life of milk (Alirezalu et al., 2020).

2.3.2.3 Effect of PEF Treatment on Vitamins and Volatile Compounds in Milk

Some researchers have studied the effect of PEF processing in milk quality parameters (Sampedro et al., 2005). These studies, in general, have concluded that PEF processing does not change the composition and quality of milk as significantly as heat treatment (Bendicho et al., 2002a). Several studies have shown that PEF has minimal effect on vitamins (Grahl & Märkl, 1996). Bendicho et al. (2002b) studied the effect of PEF processing at room and moderate temperature on water and fat-soluble vitamins in milk. The field strengths for PEF treatment used were between 18.3 to 27.1 kV/cm. The content of thiamine, riboflavin, cholecalciferol and tocopherol in milk did not change upon PEF treatment. The retention of ascorbic acid after PEF treatment was found to be higher (93.4% at 22.6 kV/cm, 400 µs treatment) than after batch (49.7%) or HTST (86.7%) pasteurization. The retention of vitamins was independent of the treatment temperature but was affected by the treatment media; skim milk retained a higher proportion of ascorbic acid than synthetic milk ultrafiltrate. Similarly, Riener, et al. (2009) reported that levels of thiamine, riboflavin, retinol and α-tocopherol remained unaffected after PEF treatment of milk at field strengths ranging from 15 to 35 kV/cm for 12.5 to 75 µs.

A recent study that combined the effect of PEF with mild heating (electric field 32 kV/cm, inlet temperatures 20–40° C, maximum outlet temperature 58° C) on whey protein isolate formulations concluded that there was no effect of PEF/heating combined processing on vitamins A and C (Schottroff et al., 2019).

Zhang et al. (2011) investigated the effects of PEF and thermal treatments on volatile compounds in bovine milk. The authors evaluated the volatile profile of the milk samples using gas chromatography/mass spectrometry (GC-MS) and used gas chromatography-olfactometry (GC-O) to characterize the volatile compounds. Raw milk was treated with 15, 20, 25 and 30 kV/cm for a total of 800 µs treatment time. The highest outlet temperature was less than 40° C. PEF-treated samples were compared with heat-pasteurized milk (75° C, 15 s).

A total of 37 volatile compounds were identified in raw bovine milk. An increase in aldehydes and ketones was reported for heat-treated milk samples. However, only aldehydes were increased in PEF-treated samples. The authors reported no significant differences in the degradation of milk triglycerides between raw and PEF-treated milk. Furthermore, no significant changes were found for acids, lactones and alcohols between raw, heat-treated or PEF-treated samples. However, 2(5H)-Furanone was only identified in PEF-treated samples. Upon GC-O analysis, no significant differences were found between heat- and PEF-pasteurized samples.

Chugh et al. (2014) compared the effect of PEF with or without tangential-flow microfiltration on the volatile composition of skim milk with HTST pasteurization. PEF treatment was conducted at 28 or 40 kV/cm for 1,122 to 2,805 µs. Nineteen volatile compounds were compared between samples. Ketones and aldehydes were significantly higher in samples heated at 95° C for 45 s while PEF treatment with or without microfiltration and heating at 75° C for 20 s did not increase the concentration of ketones and aldehydes in the milk. Significant differences in the concentrations of short-chain fatty acids (hexanoic and butanoic acids), alcohols (ethanol, propanol and toluene) and sulfur compounds (hydrogen sulfide and dimethyl sulfide) were found between PEF-treated and raw milk.

2.3.2.4 Effect of PEF Treatment on Milk Lipids, Structure of Milk Fat Globules (MFGs) and Milk Fat Globule Membranes (MFGMs)

Some recent studies have discussed the effect of PEF on milk lipids (Sharma et al., 2014c; Yang et al., 2019; Yang et al., 2021). The results of studies on the effect of PEF on milk lipids are summarized in these reviews. Some studies have suggested that the size distribution of fat globules in skim and whole milk is not affected by PEF processing (Barsotti et al., 2001; Garcia-Amezquita et al., 2009). However, a decrease in the volume mean diameter of MFG was reported by Sharma et al. (2015) at electric field intensity of 26 kV/cm applied with or without thermal pre-treatment. No change in the volume mean diameter of MFG size was reported for PEF treatment at 20 kV/cm.

Very few studies have focused on changes in the structure of MFG and MFGM due to PEF processing. These studies have shown the adsorption of caseins and whey proteins onto the MFGM after PEF treatment of milk (Sharma et al., 2015, 2016), suggesting surface damage to MFGs. Sharma et al. (2015) have been able to successfully employ transmission electron microscopy to visualize the changes in MFGM after PEF treatment. The micrographs clearly show increased casein micelle adsorption onto the MFGM after PEF treatment as compared to raw milk. It was also found that more serum proteins adsorbed onto the MFG surface when a pre-heat treatment of 55° C for 24 s was applied prior to PEF. The adsorption of proteins on MFGs' surfaces was also correlated to the intensity of the PEF treatment used (Figure 2.5).

However, this surface damage and surface fraction coverage by plasma proteins were significantly higher after the heat treatment of milk, suggesting that PEF treatment of milk is less destructive to the MFG structure. In another study (Xu et al., 2015), it was shown that PEF (37 kV/cm for 1,705 µs at 50 or 65° C) induced interactions between β-lactoglobulin and MFGM proteins, while the phospholipid composition remained unchanged. It was reported that MFGM extracted from PEF-processed cream retained its antiproliferative activity on human adenocarcinoma HT-29 cells, while this activity was lost after heat treatment of cream (Xu et al., 2015). These results suggested that the PEF treatment can preserve the structural and biological functions of MFGM.

There are very few studies on the effect of PEF on the free fatty acid (FFA) content and profile of milk after treatment. It can be hypothesized that damage of MFGM by PEF can lead to lipolysis by the action of indigenous milk lipase enzyme, lipoprotein lipase (LPL) and/or microbial lipases. However, as discussed in section 2.3.2.2, depending on the processing conditions, lipases can be inactivated by PEF (Bendicho et al., 2002c; Riener et al., 2009; Sharma et al., 2014b). Hence, the effect of PEF on the FFA profile will be a balance between its effect on MFGM and lipases present in milk.

In a continuous-flow bench-scale PEF system, at an electric field strength of 35.5 kV/cm for 300 or 1,000 µs, Odriozola-Serrano et al. (2006) reported that the FFA content of PEF-processed milk was similar to that of unprocessed milk immediately after processing. However, during storage, the FFA content of milk changed significantly. While the FFA content increased significantly in a sample treated for 300 µs with PEF after 11 days of storage, the increase in FFA content of a 1,000 µs treated sample was similar to that of a heat-pasteurized sample. The authors suggested that this change in FFA content was perhaps due to the spoilage of milk by lipase-secreting

2.3 Application of PEF to Milk Processing 27

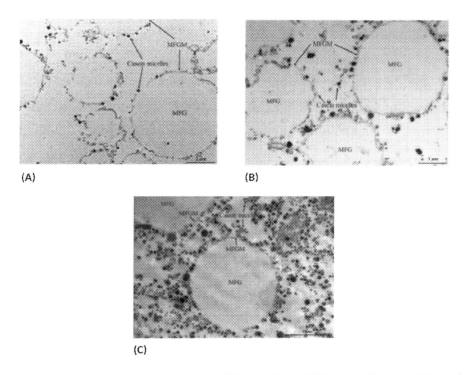

Figure 2.5 Transmission electron micrographs of (a) raw milk at 5,800× magnification with a scale bar of 2 μm; (b) pulsed electric field (PEF)-treated milk (E: 20 kV cm−1) at 13,500× magnification with a scale bar of 1 μm; (c) milk pre-heated to 55° C for 24 s and followed by PEF (E: 20 kV cm−1) at 13,500× magnification with a scale bar of 1 μm. Source: directly taken from Figure 6 from Sharma et al. (2015) with permission.

microorganisms. McAuley et al. (2016) suggested that the concentration of short-chain fatty acids (butanoic, hexanoic and octanoic acids) was much higher in raw milk than in PEF-treated milk during 13-day storage at 4° C. This suggested lower activity of LPL and microbial lipases in PEF-treated milk.

In a recent study, Yang et al. (2021) reported how the milk fat globule size alters the effect of PEF treatment on milk fat globule structure and the fatty acid composition of milk. It was also reported that PEF treatment significantly altered the fatty acid composition of the milk. There was a significant increase in the proportion of short- and medium-chain fatty acids after PEF treatment while the proportion of long-chain fatty acids (C14 to C16:1) decreased. PEF also resulted in the appearance of fatty acids in treated milk that were not present in the raw milk before treatment. These changes could be attributed to a lower inactivation of lipases in this study due to lower electric field strength (9 or 16 kV/cm). MFGs in samples containing larger globules showed a higher tendency toward clumping into larger-sized globules after PEF treatment. Confocal microscopy also showed that PEF treatment resulted in adsorption of the smallest MFG, phospholipid fragments and/or milk proteins onto the MFG surface, confirming the results of Sharma et al. (2015). The study also reported that after PEF treatment, α-lactalbumin (α-la) and β-lactoglobulin (β-lg) from milk serum preferentially adsorbed onto the smaller MFGs. Also, the natural proteins in the MFGM of the smaller MFGs were more significantly affected after PEF treatment.

2.3.2.5 Effect of PEF Treatment on Milk Proteins
2.3.2.5.1 Caseins
Some researchers have studied the effect of PEF treatment on the properties and integrity of casein micelles (Sharma et al., 2014c; Yang et al., 2019). However, inconsistent results have been reported, perhaps due to the differences in PEF equipment and test conditions used. While studying PEF treatment of skim milk, Floury et al. (2006) suggested that PEF treatment at 45–55 kV/cm for 2.1–3.5 µs reduced the casein micelle size while maintaining the temperature of processing below 50° C, while Liu et al. (2015) showed that there was no change in casein micelle size below pH 7.5 when treated at 49 kV/cm for 19.36 µs. Dissociation of casein micelles and a concurrent decrease in casein micelle size were observed in samples at pH ≥ 7.5. It was also suggested that the main reason for the change in the size of casein micelles was ohmic heating during the PEF treatment. Michalac et al. (2003), however, did not report any change in casein micelles when treating milk at 35 kV/cm for 188 µs. It has also been shown that the effect of PEF on casein micelle size is dependent on the time elapsed between treatment and measurement. Hemar et al. (2011) reported that the casein micelle size reverted to the original size in untreated milk after overnight cold storage.

2.3.2.5.2 Whey Proteins
Several studies have analyzed the effect of PEF treatment on whey proteins in buffers as well as in milk. Being globular proteins, whey proteins are more susceptible to changes than caseins.

Some early studies suggested that PEF treatment did not affect the whey proteins in milk (Dunn, 1996; Grahl & Märkl, 1996; Michalac et al., 2003). While studying a high-intensity PEF process (35.5 kV/cm for 1,000 or 300 µs), Odriozola-Serrano et al. (2006) found that the PEF process can affect the whey proteins in milk. They reported that the retention of major whey proteins, serum albumin, α-la and β-lg, was between 79.9% and 60% with α-la being most affected. However, the decrease in retention of the whey proteins after PEF treatment was similar to that after heat treatment of milk. Sharma et al. (2016) compared the differential scanning calorimetric scans of untreated and PEF-treated milk (20 or 26 kV/cm for 34 µs). It was found that after PEF treatment new protein–protein complexes were formed in milk, perhaps resulting from interactions between MFGM proteins and whey proteins. On the contrary, Liu et al. (2015) showed that PEF treatment at 49 kV/cm for 19.36 µs did not affect the whey proteins even though the treatment temperature reached up to 70° C. While studying a reconstituted whey protein isolate system, Sui et al. (2011) also showed that PEF treatment of 35 kV/cm for 19.2 µs did not affect the physicochemical properties of the whey proteins. Similarly, Schottroff et al. (2019) showed that PEF of 2% whey protein isolate solution did not influence the immunoglobulins in the sample.

Several studies on the effect of PEF on whey proteins have been recently reviewed (Yang et al., 2019). It is suggested that factors such as electric field intensity, number of pulses, the concentration of proteins, pH and temperature can govern the impact of PEF processing on whey proteins in milk.

2.4 CONCLUSIONS
Milk and milk products are highly regarded for their nutritional value and play an indispensable role in meeting the nutritional requirements of human beings of all ages from neonates to adults to the elderly. The current widely applied industrial process of thermal pasteurization, while effectively decontaminating and extending the microbial shelf-life of milk, can have

adverse effects on its essential properties including flavor, color, density and valuable nutrient compounds as well as bioactivity. Pulsed power technology generates high-voltage pulses and delivers energy through short pulses in the microseconds to nanoseconds (μs–ns) range. Bioelectrics is an emerging field that applies pulsed power energy to biomedical and food processing. The application of pulsed power in food processing is increasing for several applications such as nutrient extraction, food processing, microbial inactivation (and hence, shelf-life extension) and growth stimulation. A virtue of pulsed power technology is that the electrical parameters of the pulses (such as amplitude, pulse width, frequency, rise time and falling time) can be manipulated to adjust the delivered energy and applied electric field that suit the related application. Microbial inactivation using pulsed power is achieved by increased permeabilization and/or formation of pores in the cellular membrane. However, the efficiency of pulsed power treatment in microbial inactivation is dependent on several pulse parameters, treatment chamber configuration, temperature and the type of material (load) being treated. Hence, for an efficient treatment process, the pulsed power system (generated pulses and treatment chambers) must be studied and designed at two different levels, the system level and the load level. While several studies are available detailing the effect of pulsed power treatment on microbial inactivation in milk, the results of its effect on milk components are often conflicting. A major reason for this is that several different setups and treatment conditions are used. Some studies accurately measure and report all important parameters, such as processing temperature, while others do not. To achieve the industrial-level application of pulsed electric fields in milk processing, future research needs to focus on consistency in designing and reporting study parameters and scale-up trials.

REFERENCES

Akiyama, H., & Heller, R. (2017). *Bioelectrics*. Japan: Springer.

Alirezalu, K., Munekata, P. E., Parniakov, O., Barba, F. J., Witt, J., Toepfl, S., Wiktor, A., & Lorenzo, J. M. (2020). Pulsed electric field and mild heating for milk processing: A review on recent advances. *Journal of the Science of Food & Agriculture*, *100*(1), 16–24.

Ashokkumar, M., Bhaskaracharya, R., Kentish, S., Lee, J., Palmer, M., & Zisu, B. (2010). The ultrasonic processing of dairy products—An overview. *Dairy Science & Technology*, *90*(2–3), 147–168.

Barbosa-Cánovas, G., & Bermúdez-Aguirre, D. (2010). Pasteurization of milk with pulsed electric fields. In: M. W. Griffiths (Ed.), *Improving the Safety and Quality of Milk*, Vol. 17, (pp. 400–419) Cambridge, UK: Woodhead Publishing Limited.

Barsotti, L., Dumay, E., Mu, T. H., Fernandez Diaz, M. D., & Cheftel, J. C. (2001). Effects of high voltage electric pulses on protein-based food constituents and structures. *Trends in Food Science & Technology*, *12*(3–4), 136–144.

Beebe, S. J., & Schoenbach, K. H. (2005). Nanosecond pulsed electric fields: A new stimulus to activate intracellular signaling. *Journal of Biomedicine & Biotechnology*, 2005(4), 297–300.

Bendicho, S., Barbosa-Cánovas, G. V., & Martín, O. (2002a). Milk processing by high intensity pulsed electric fields. *Trends in Food Science & Technology*, *13*(6–7), 195–204.

Bendicho, S., Espachs, A., Arántegui, J., & Martín, O. (2002b). Effect of high intensity pulsed electric fields and heat treatments on vitamins of milk. *The Journal of Dairy Research*, *69*(1), 113.

Bendicho, S., Estela, C., Giner, J., Barbosa-Cánovas, G., & Martin, O. (2002c). Effects of high intensity pulsed electric field and thermal treatments on a lipase from *Pseudomonas fluorescens*. *Journal of Dairy Science*, *85*(1), 19–27.

Bermúdez-Aguirre, D., & Barbosa-Cánovas, G. V. (2011). Power ultrasound to process dairy products. In: F. Hao, B.-C. Gustavo & W. Jochen (Eds.), *Ultrasound Technologies for Food and Bioprocessing*, (pp. 445–465) Spain: Springer.

Bertucco, A., & Vetter, G. (2001). *High Pressure Process Technology: Fundamentals and Applications.* Amsterdam, The Netherlands: Elsevier Science B.V.

Bochkov, D. V., Gnedin, I. N., Vasiliev, G. M., Vasetskiy, V. A., & Zhdanok, S. A. (2012). High voltage pulse generator based on TPI-thyratron for pulsed electric field milk processing. In: IEEE International Power Modulator and High Voltage Conference (IPMHVC), (pp. 98–101): IEEE.

Buckow, R., Semrau, J., Sui, Q., Wan, J., & Knoerzer, K. (2012). Numerical evaluation of lactoperoxidase inactivation during continuous pulsed electric field processing. *Biotechnology Progress, 28*(5), 1363–1375.

Calderón-Miranda, M. L., Barbosa-Cánovas, G. V., & Swanson, B. G. (1999). Transmission electron microscopy of *Listeria innocua* treated by pulsed electric fields and nisin in skimmed milk. *International Journal of Food Microbiology, 51*(1), 31–38.

Castro, A., Swanson, B., Barbosa-Canovas, G., & Dunker, A. (2001a). Pulsed electric field denaturation of bovine alkaline phosphatase. In: G. V. Barbosa-Cánovas & Q. H. Zhang (Eds.), *Pulsed Electric Fields in Food Processing: Fundamental Aspects and Applications,* (pp. 83–103) Florida, USA: CRC Press.

Castro, A., Swanson, B., Barbosa-Cánovas, G., & Zhang, Q. H. (2001b). Pulsed electric field modification of milk alkaline phosphatase activity. In: G. V. Barbosa-Cánovas & Q. H. Zhang (Eds.), *Pulsed Electric Fields in Food Processing: Fundamental Aspects and Applications,* (pp. 65–83) Florida, USA: CRC Press.

Chugh, A., Khanal, D., Walkling-Ribeiro, M., Corredig, M., Duizer, L., & Griffiths, M. W. (2014). Change in color and volatile composition of skim milk processed with pulsed electric field and microfiltration treatments or heat pasteurization. *Foods, 3*(2), 250–268.

Claeys, W. L., Cardoen, S., Daube, G., De Block, J., Dewettinck, K., Dierick, K., De Zutter, L., Huyghebaert, A., Imberechts, H., Thiange, P., Vandenplas, Y., & Herman, L. (2013). Raw or heated cow milk consumption: Review of risks and benefits. *Food Control, 31*(1), 251–262.

D'amico, D. J., Silk, T. M., Wu, J., & Guo, M. (2006). Inactivation of microorganisms in milk and apple cider treated with ultrasound. *Journal of Food Protection, 69*(3), 556–563.

Dunn, J. (1996). Pulsed light and pulsed electric field for foods and eggs. *Poultry Science, 75*(9), 1133–1136.

Ertugay, M., Yuksel, Y., & Sengul, M. (2003). The effect of ultrasound on lactoperoxidase and alkaline phosphatase enzymes from milk. *Milchwissenschaft, 58*(11/12), 593–595.

Floury, J., Grosset, N., Leconte, N., Pasco, M., Madec, M.-N., & Jeantet, R. (2006). Continuous raw skim milk processing by pulsed electric field at non-lethal temperature: Effect on microbial inactivation and functional properties. *Le Lait, 86*(1), 43–57.

Fox, P. F., McSweeney, P. L., & Paul, L. (1998). *Dairy Chemistry and Biochemistry.* London: Blackie Academic & Professional.

Garcia-Amezquita, L. E., Primo-Mora, A. R., Barbosa-Cánovas, G. V., & Sepulveda, D. R. (2009). Effect of nonthermal technologies on the native size distribution of fat globules in bovine cheese-making milk. *Innovative Food Science & Emerging Technologies, 10*(4), 491–494.

García, D., Gómez, N., Mañas, P., Condón, S., Raso, J., & Pagán, R. (2005a). Occurrence of sublethal injury after pulsed electric fields depending on the micro-organism, the treatment medium ph and the intensity of the treatment investigated. *Journal of Applied Microbiology, 99*(1), 94–104.

García, D., Gómez, N., Mañas, P., Raso, J., & Pagán, R. (2007). Pulsed electric fields cause bacterial envelopes permeabilization depending on the treatment intensity, the treatment medium pH and the microorganism investigated. *International Journal of Food Microbiology, 113*(2), 219–227.

García, D., Gómez, N., Raso, J., & Pagán, R. (2005b). Bacterial resistance after pulsed electric fields depending on the treatment medium pH. *Innovative Food Science & Emerging Technologies, 6*(4), 388–395.

Ghasemi, N., Zhang, J., Zare, F., & Bansal, N. (2020). Real-time method for rapid microbial assessment of bovine milk treated by nanosecond pulsed electric field. *IEEE Transactions on Plasma Science, 48*(12), 4221–4227.

Gianulis, E. C., Casciola, M., Xiao, S., Pakhomova, O. N., & Pakhomov, A. G. (2018). Electropermeabilization by uni-or bipolar nanosecond electric pulses: The impact of extracellular conductivity. *Bioelectrochemistry, 119,* 10–19.

Grahl, T., & Märkl, H. (1996). Killing of microorganisms by pulsed electric fields. *Applied Microbiology & Biotechnology, 45*(1–2), 148–157.

Guerrero-Beltrán, J. Á., Sepulveda, D. R., Góngora-Nieto, M. M., Swanson, B., & Barbosa-Cánovas, G. V. (2010). Milk thermization by pulsed electric fields (PEF) and electrically induced heat. *Journal of Food Engineering, 100*(1), 56–60.

References

Harvey, E. N., & Loomis, A. L. (1929). The destruction of luminous bacteria by high frequency sound waves. *Journal of Bacteriology*, 17(5), 373.

Heinz, V., Álvarez, I., Angersbach, A., & Knorr, D. (2001). Preservation of liquid foods by high intensity pulsed electric fields—Basic concepts for process design. *Trends in Food Science & Technology*, 12(3–4), 103–111.

Hemar, Y., Augustin, M. A., Cheng, L., Sanguansri, P., Swiergon, P., & Wan, J. (2011). The effect of pulsed electric field processing on particle size and viscosity of milk and milk concentrates. *Milchwissenschaft*, 66, 126–128.

Ho, S. Y., & Mittal, G. S. (2000). High voltage pulsed electrical field for liquid food pasteurization. *Food Reviews International*, 16(4), 395–434.

Ho, S. Y., Mittal, G. S., & Cross, J. D. (1997). Effects of high field electric pulses on the activity of selected enzymes. *Journal of Food Engineering*, 31(1), 69–84.

Jäger, H., & Knorr, D. (2017). Pulsed electric fields treatment in food technology: Challenges and opportunities. In: Damijan Miklavčič (ed.) *Handbook of Electroporation*, (pp. 2657–2680): Cham: Springer.

Jayaram, S., Castle, G., & Margaritis, A. (1993). The effects of high field DC pulse and liquid medium conductivity on survivability of *Lactobacillus brevis*. *Applied Microbiology & Biotechnology*, 40(1), 117–122.

Kotnik, T., Bobanović, F., & Miklavčič, D. (1997). Sensitivity of transmembrane voltage induced by applied electric fields—A theoretical analysis. *Bioelectrochemistry & Bioenergetics*, 43(2), 285–291.

Kotnik, T., Rems, L., Tarek, M., & Miklavčič, D. (2019). Membrane electroporation and electropermeabilization: Mechanisms and models. *Annual Review of Biophysics*, 48, 63–91.

Lawton, B., & Pethig, R. (1993). Determining the fat content of milk and cream using AC conductivity measurements. *Measurement Science & Technology*, 4(1), 38.

Lee, G. J., Han, B. K., Choi, H. J., Kang, S. H., Baick, S. C., & Lee, D.-U. (2015). Inactivation of *Escherichia coli*, *Saccharomyces cerevisiae*, and *Lactobacillus brevis* in low-fat milk by pulsed electric field treatment: A pilot-scale study. *Korean Journal for Food Science of Animal Resources*, 35(6), 800.

Liu, Z., Hemar, Y., Tan, S., Sanguansri, P., Niere, J., Buckow, R., & Augustin, M. A. (2015). Pulsed electric field treatment of reconstituted skim milks at alkaline pH or with added EDTA. *Journal of Food Engineering*, 144, 112–118.

Mabrook, M., & Petty, M. (2003). Effect of composition on the electrical conductance of milk. *Journal of Food Engineering*, 60(3), 321–325.

McAuley, C. M., Singh, T. K., Haro-Maza, J. F., Williams, R., & Buckow, R. (2016). Microbiological and physicochemical stability of raw, pasteurised or pulsed electric field-treated milk. *Innovative Food Science & Emerging Technologies*, 38, 365–373.

McClements, D. J. (1995). Advances in the application of ultrasound in food analysis and processing. *Trends in Food Science & Technology*, 6(9), 293–299.

Mertens, B. (1995). Hydrostatic pressure treatment of food: Equipment and processing. In: G. W. Gould (Ed.), *New Methods of Food Preservation*, (pp. 135–158): Dordrecht, The Netherlands: Springer Science +Business Media B.V.

Michalac, S., Alvarez, V., Ji, T., & Zhang, Q. H. (2003). Inactivation of selected microorganisms and properties of pulsed electric field processed milk. *Journal of Food Processing & Preservation*, 27(2), 137–151.

Mújica-Paz, H., Valdez-Fragoso, A., Samson, C. T., Welti-Chanes, J., & Torres, J. A. (2011). High-pressure processing technologies for the pasteurization and sterilization of foods. *Food & Bioprocess Technology*, 4(6), 969–985.

Odriozola-Serrano, I., Bendicho-Porta, S., & Martín-Belloso, O. (2006). Comparative study on shelf life of whole milk processed by high-intensity pulsed electric field or heat treatment. *Journal of Dairy Science*, 89(3), 905–911.

OECD/FAO (2020). *OECD-FAO Agricultural Outlook*: Paris: FAO; Rome: OECD Publishing, (pp. 2020–2029).

Ohshima, T., Tamura, T., & Sato, M. (2007). Influence of pulsed electric field on various enzyme activities. *Journal of Electrostatics*, 65(3), 156–161.

Ohshima, T., Tanino, T., Kameda, T., & Harashima, H. (2016). Engineering of operation condition in milk pasteurization with PEF treatment. *Food Control*, 68, 297–302.

Pagán, R., Mañas, P., Raso, J., & Condón, S. (1999). Bacterial resistance to ultrasonic waves under pressure at nonlethal (manosonication) and lethal (manothermosonication) temperatures. *Applied & Environmental Microbiology*, 65(1), 297–300.

Pakhomov, A. G., Grigoryev, S., Semenov, I., Casciola, M., Jiang, C., & Xiao, S. (2018). The second phase of bipolar, nanosecond-range electric pulses determines the electroporation efficiency. *Bioelectrochemistry*, *122*, 123–133.

Picart, L., Dumay, E., & Cheftel, J. C. (2002). Inactivation of *Listeria innocua* in dairy fluids by pulsed electric fields: Influence of electric parameters and food composition. *Innovative Food Science & Emerging Technologies*, *3*(4), 357–369.

Poojary, M., Roohinejad, S., Koubaa, M., Barba, F., Passamonti, P., Jambrak, A. R., Oey, I., & Greiner, R. (2016). Impact of pulsed electric fields on enzymes. In: D. Miklavčič (Ed.), *Handbook of Electroporation*, (pp. 1–21): Springer International Publishing.

Qin, B.-L., Barbosa-Canovas, G. V., Swanson, B. G., Pedrow, P. D., & Olsen, R. G. (1998). Inactivating microorganisms using a pulsed electric field continuous treatment system. *IEEE Transactions on Industry Applications*, *34*(1), 43–50.

Qin, B. L., Pothakamury, U. R., Barbosa-Cánovas, G. V., & Swanson, B. G. (1996). Nonthermal pasteurization of liquid foods using high-intensity pulsed electric fields. *Critical Reviews in Food Science & Nutrition*, *36*(6), 603–627.

Rebersek, M., Miklavčič, D., Bertacchini, C., & Sack, M. (2014). Cell membrane electroporation-Part 3: The equipment. *IEEE Electrical Insulation Magazine*, *30*(3), 8–18.

Reina, L. D., Jin, Z. T., Zhang, Q. H., & Yousef, A. E. (1998). Inactivation of *Listeria monocytogenes* in milk by pulsed electric field. *Journal of Food Protection*, *61*(9), 1203–1206.

Riener, J., Noci, F., Cronin, D. A., Morgan, D. J., & Lyng, J. G. (2009). Effect of high intensity pulsed electric fields on enzymes and vitamins in bovine raw milk. *International Journal of Dairy Technology*, *62*(1), 1–6.

Sampedro, F., & Rodrigo, D. (2015). Pulsed electric fields (PEF) processing of milk and dairy products. In: D. Nivedita & Peggy M. Tomasula (Eds.), *Emerging Dairy Processing Technologies: Opportunities for the Dairy Industry*, (pp. 115–148): West Sussex, UK: John Wiley & Sons, Ltd.

Sampedro, F., Rodrigo, M., Martínez, A., Rodrigo, D., & Barbosa-Cánovas, G. (2005). Quality and safety aspects of PEF application in milk and milk products. *Critical Reviews in Food Science & Nutrition*, *45*(1), 25–47.

Saulis, G. (2010). Electroporation of cell membranes: The fundamental effects of pulsed electric fields in food processing. *Food Engineering Reviews*, *2*(2), 52–73.

Schoenbach, K. H., Katsuki, S., Stark, R. H., Buescher, E. S., & Beebe, S. J. (2002). Bioelectrics-new applications for pulsed power technology. *IEEE Transactions on Plasma Science*, *30*(1), 293–300.

Schottroff, F., Gratz, M., Krottenthaler, A., Johnson, N. B., Bédard, M. F., & Jaeger, H. (2019). Pulsed electric field preservation of liquid whey protein formulations-Influence of process parameters, pH, and protein content on the inactivation of *Listeria innocua* and the retention of bioactive ingredients. *Journal of Food Engineering*, *243*, 142–152.

Schottroff, F., Johnson, K., Johnson, N. B., Bédard, M. F., & Jaeger, H. (2020). Challenges and limitations for the decontamination of high solids protein solutions at neutral pH using pulsed electric fields. *Journal of Food Engineering*, *268*, 109737.

Semenov, I., Xiao, S., Pakhomova, O. N., & Pakhomov, A. G. (2013). Recruitment of the intracellular Ca^{2+} by ultrashort electric stimuli: The impact of pulse duration. *Cell Calcium*, *54*(3), 145–150.

Shamsi, K., Versteeg, C., Sherkat, F., & Wan, J. (2008). Alkaline phosphatase and microbial inactivation by pulsed electric field in bovine milk. *Innovative Food Science & Emerging Technologies*, *9*(2), 217–223.

Sharma, P., Bremer, P., Oey, I., & Everett, D. W. (2014a). Bacterial inactivation in whole milk using pulsed electric field processing. *International Dairy Journal*, *35*(1), 49–56.

Sharma, P., Oey, I., Bremer, P., & Everett, D. W. (2014b). Reduction of bacterial counts and inactivation of enzymes in bovine whole milk using pulsed electric fields. *International Dairy Journal*, *39*(1), 146–156.

Sharma, P., Oey, I., & Everett, D. W. (2014c). Effect of pulsed electric field processing on the functional properties of bovine milk. *Trends in Food Science & Technology*, *35*(2), 87–101.

Sharma, P., Oey, I., & Everett, D. W. (2015). Interfacial properties and transmission electron microscopy revealing damage to the milk fat globule system after pulsed electric field treatment. *Food Hydrocolloids*, *47*, 99–107.

Sharma, P., Oey, I., & Everett, D. W. (2016). Thermal properties of milk fat, xanthine oxidase, caseins and whey proteins in pulsed electric field-treated bovine whole milk. *Food Chemistry*, *207*, 34–42.

Shin, J.-K., Jung, K.-J., Pyun, Y.-R., & Chun, M.-S. (2007). Application of pulsed electric fields with square wave pulse to milk inoculated with *E. coli*, *P. fluorescens*, and *B. stearothermophilus*. *Food Science & Biotechnology*, *16*(6), 1082–1084.

Siemer, C., Toepfl, S., & Heinz, V. (2014). Inactivation of *Bacillus subtilis* spores by pulsed electric fields (PEF) in combination with thermal energy II. Modeling thermal inactivation of *B. subtilis* spores during PEF processing in combination with thermal energy. *Food Control*, *39*, 244–250.

Smelt, J. P. P. M. (1998). Recent advances in the microbiology of high pressure processing. *Trends in Food Science & Technology*, *9*(4), 152–158.

Smith, K., Mittal, G., & Griffiths, M. (2002). Pasteurization of milk using pulsed electrical field and antimicrobials. *Journal of Food Science*, *67*(6), 2304–2308.

Sobrino-López, A., Raybaudi-Massilia, R., & Martín-Belloso, O. (2006). High-intensity pulsed electric field variables affecting *Staphylococcus aureus* inoculated in milk. *Journal of Dairy Science*, *89*(10), 3739–3748.

Spilimbergo, S., Dehghani, F., Bertucco, A., & Foster, N. R. (2003). Inactivation of bacteria and spores by pulse electric field and high pressure CO_2 at low temperature. *Biotechnology & Bioengineering*, *82*(1), 118–125.

Sui, Q., Roginski, H., Williams, R. P. W., Versteeg, C., & Wan, J. (2011). Effect of pulsed electric field and thermal treatment on the physicochemical and functional properties of whey protein isolate. *International Dairy Journal*, *21*(4), 206–213.

Syed, Q. A., Ishaq, A., Rahman, U., Aslam, S., & Shukat, R. (2017). Pulsed electric field technology in food preservation: A review. *Journal of Nutritional Health & Food Engineering*, *6*(6), 168–172.

Toepfl, S., Siemer, C., Saldaña-Navarro, G., & Heinz, V. (2005). Overview of pulsed electric fields processing for food. In: D.-W. Sun (Ed.), *Emerging Technologies for Food Processing*, (pp. 93–114) California, CA: Academic Press.

Vaessen, E. M. J., Timmermans, R. A. H., Tempelaars, M. H., Schutyser, M. A. I., & den Besten, H. M. W. (2019). Reversibility of membrane permeabilization upon pulsed electric field treatment in *Lactobacillus plantarum* WCFS1. *Scientific Reports*, *9*(1), 19990–19911.

Van Loey, A., Verachtert, B., & Hendrickx, M. (2001). Effects of high electric field pulses on enzymes. *Trends in Food Science & Technology*, *12*(3–4), 94–102.

Vega-Mercado, H., Pothakamury, U. R., Chang, F.-J., Barbosa-Cánovas, G. V., & Swanson, B. G. (1996). Inactivation of *Escherichia coli* by combining pH, ionic strength and pulsed electric fields hurdles. *Food Research International*, *29*(2), 117–121.

Vega-Mercado, H., Powers, J. R., Barbosa-Cánovas, G. V., & Swanson, B. G. (1995). Plasmin inactivation with pulsed electric fields. *Journal of Food Science*, *60*(5), 1143–1146.

Vercet, A., Burgos, J., Crelier, S., & Lopez-Buesa, P. (2001). Inactivation of proteases and lipases by ultrasound. *Innovative Food Science & Emerging Technologies*, *2*(2), 139–150.

Villamiel, M., & de Jong, P. (2000). Influence of high-intensity ultrasound and heat treatment in continuous flow on fat, proteins, and native enzymes of milk. *Journal of Agricultural & Food Chemistry*, *48*(2), 472–478.

Walstra, P., Geurts, T. J., Noomen, A., Jellema, A., & van Boekel, M. A. J. S. (1999). *Dairy Technology: Principles of Milk Properties and Processes*. Florida, USA: CRC Press.

Walter, L., Knight, G., Ng, S. Y., & Buckow, R. (2016). Kinetic models for pulsed electric field and thermal inactivation of *Escherichia coli* and *Pseudomonas fluorescens* in whole milk. *International Dairy Journal*, *57*, 7–14.

Wong, N. P. (2012). *Fundamentals of Dairy Chemistry*. New York, NY: Springer Science & Business Media.

Xu, S., Walkling-Ribeiro, M., Griffiths, M. W., & Corredig, M. (2015). Pulsed electric field processing preserves the antiproliferative activity of the milk fat globule membrane on colon carcinoma cells. *Journal of Dairy Science*, *98*(5), 2867–2874.

Yang, S., Liu, G., Qin, Z., Munk, D., Otte, J., & Ahrné, L. (2019). Effects of pulsed electric fields on food constituents, microstructure and sensorial attributes of food products. In: S. Roohinejad, M. Koubaa, R. Greiner & K. Mallikarjunan (Eds.), *Effect of Emerging Processing Methods on the Food Quality*, (pp. 27–67) Switzerland: Springer.

Yang, S., Suwal, S., Andersen, U., Otte, J., & Ahrné, L. (2021). Effects of pulsed electric field on fat globule structure, lipase activity, and fatty acid composition in raw milk and milk with different fat globule sizes. *Innovative Food Science & Emerging Technologies*, *67*, 102548.

Yu, M., & Lin, H. (2014). Quantification of propidium iodide delivery with millisecond electric pulses: A model study. *ArXiv Preprint, 1401*, 6954.

Zhang, S., Yang, R., Zhao, W., Hua, X., Zhang, W., & Zhang, Z. (2011). Influence of pulsed electric field treatments on the volatile compounds of milk in comparison with pasteurized processing. *Journal of Food Science, 76*(1), C127–C132.

Zhao, W., Yang, R., & Zhang, H. Q. (2012). Recent advances in the action of pulsed electric fields on enzymes and food component proteins. *Trends in Food Science & Technology, 27*(2), 83–96.

Żywica, R., Banach, J. K., & Kiełczewska, K. (2012). An attempt of applying the electrical properties for the evaluation of milk fat content of raw milk. *Journal of Food Engineering, 111*(2), 420–424.

Chapter 3

High Hydrostatic Pressure Processing for Dairy Products

M. Selvamuthukumaran, Nilesh Nirmal and Sajid Maqsood

CONTENTS

3.1	High-Pressure Processing/High Hydrostatic Pressure Processing	35
	3.1.1 Introduction	35
3.2	Market Milk	36
3.3	Milkshakes	36
	3.3.1 Effect of High-Pressure Processing on Total Phenol Content and Antioxidative Capacity of Milkshakes	37
	3.3.2 Effect of High-Pressure Processing on Microbial Count of a Chokeberry-Incorporated Milkshake	37
3.4	Yogurt	38
3.5	Cheese	39
3.6	Conclusion	39
References		39

3.1 HIGH-PRESSURE PROCESSING/HIGH HYDROSTATIC PRESSURE PROCESSING

3.1.1 Introduction

Dairy products are usually subjected to various thermal treatments like high-temperature short-time pasteurization, ultra-high temperature pasteurization or low-temperature long-time pasteurization for enhancing the shelf stability of the product (Barraquio, 2014).

The application of traditional heat processing techniques will have some destructive effects on the nutritional profile of dairy products. The thermal treatment exposure of dairy products may lead to the denaturation of whey protein by 5 to 15%, the destruction of water-soluble vitamins, an increase in the allergenic milk protein properties and structural disturbances in casein micelles (Barraquio, 2014; Tamime, 2014; Bogahawaththa et al., 2018).

Therefore, high-pressure processing or high hydrostatic pressure processing is one of the substitutes for traditional heat treatment techniques in order to destroy foodborne pathogens; it reduces the nutrient loss mainly of water-soluble vitamins, so that food product freshness can be maintained to a greater extent (Considine et al., 2008; Wang et al., 2015) (Figure 3.1). In

DOI: 10.1201/9781003138716-3

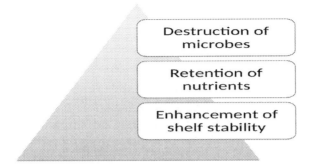

Figure 3.1 Advantages of using high-pressure processing for dairy foods.

high-pressure processing, the pressure will stay in contact with the product for a particular holding period in order to trigger the microbial spores' destruction and as well as the non-covalent bond formation of food to gelatinize enzymes (Wang et al., 2015; Chawla et al., 2011; Elamin et al., 2015). The treatment of high pressure can result in the destruction of microbial cells so that the shelf-life of such dairy products can be significantly enhanced.

3.2 MARKET MILK

Tan et al. (2020) adopted high-pressure processing technology for cow as well as goat milk on a commercial basis in comparison with traditional heat pasteurization techniques. Their studies showed that treated milk doesn't exhibit physicochemical changes except for pH. The high-pressure processed goat milk exhibited enhanced pH for both pressure-treated and pasteurized samples. The pressure-treated samples exhibited microbial stability of 22 days when they were stored at 8° C without enhancing microorganisms like coliform, yeast, mold, mesophilic spores or *B. cereus*, with a noticeable increase in the total microbial count and as well as the psychotropic bacterial count. The pressure-processed cow milk sample projected more physicochemical stability when compared to goat milk, justified by insignificant changes of acidity levels compared to processed goat milk, which exhibited a 0.04% acidity increase during storage.

3.3 MILKSHAKES

High-pressure processing can preserve food as well as enhancing nutritional bioavailability and bioaccessibility and also significantly enhancing the antioxidant activity ascribed to microstructure modifications (Vázquez-Gutiérrez et al., 2013). It also retains the product's organoleptic characteristics (San Martín et al., 2002); their application helps to inactivate vegetative cells when the products are subjected to pressure (Welti-Chanes et al., 2005).

Elena Diez-Sánchez et al. (2020) developed a milkshake by fortifying polyphenol-rich pomace obtained from chokeberries. They used high-pressure processing for the extraction of polyphenol from chokeberry pomace. They enriched the milk with various proportions of chokeberry pomace and they further studied the application of high-pressure processing, i.e. pressure and time period's effects on constituents like total phenol content, antioxidative capacity and the inactivation of microbes.

3.3.1 Effect of High-Pressure Processing on Total Phenol Content and Antioxidative Capacity of Milkshakes

The total phenol content for a 10% incorporated chokeberry pomace sample exhibited lower content of 121 mg/100 ml when compared to a high-pressure processed sample, which recorded total phenol content of 134 mg/100 ml with pressure application of 500 MPa for a time period of 10 min (Table 3.1a). The lower chokeberry incorporated sample, i.e. 2.5%, exhibited 53 mg/100 ml. The 2.5% chokeberry pomace incorporation in a milkshake (untreated one) exhibited an antioxidative capacity of 6 μmol Trolox/mL when compared to a pressure-treated sample, i.e. 7 μmol Trolox/mL at 500 MPa for a time period of 10 min (Table. 3.1b). Therefore, it is justified to conclude that the pressure-treated samples significantly enhanced the total phenol content as well as the extraction efficiency of antioxidative constituents (Tokusoglu, 2016).

The plant tissue breakdown and disintegration of the cell wall after pressure treatment led to enhancing the phenolic content and antioxidative capacity ascribed to contents that might have leached out from pomace cells to the milk blend (Vázquez-Gutiérrez et al., 2013; Corrales et al., 2008; Gao et al., 2016; Hernández-Carrión et al., 2014). The effect was highly noticed when samples were subjected to a pressure treatment of 220 MPa, where proteins get unfolded and the occurrence of interface separation is quite possible.

3.3.2 Effect of High-Pressure Processing on Microbial Count of a Chokeberry-Incorporated Milkshake

The exposure of a chokeberry milkshake to high-pressure processing had resulted in the inactivation of microbes ascribed to microbial cell-induced changes. It was explained by several researchers that the product's high-pressure treatment exposure will lead to alterations of the cell membrane and bring protein effects as well as microbial genetic mechanism effects (Welti-Chanes et al., 2005; Patterson, 2005; Ritz et al., 2002). The incorporation of 2.5% pomace with

TABLE 3.1 EFFECT OF HIGH-PRESSURE PROCESSING* ON RETENTION OF TOTAL PHENOL CONTENT AND ANTIOXIDATIVE CONSTITUENTS IN CHOKEBERRY-INCORPORATED MILKSHAKES**

(a)

Constituents	0	200	350	500
Total phenol content (mg/100 ml)	121	133	125	134
Antioxidative capacity (μmol Trolox/mL)	15	14	17	15

Notes:
* Treatment time: 10 min.
** Chokeberry incorporation at 10%.

(b)

Constituents	0	200	350	500
Total phenol content (mg/100 ml)	53	51	42	58
Antioxidative capacity (μmol Trolox/mL)	6	5	5	7

Notes:
* Treatment time: 10 min.
** Chokeberry incorporation at 2.5%.

200 MPa high-pressure treatments for a time period of 1 min as well as 10 min doesn't bring any significant inactivation of microbial cells. *L. monocytogenes* inactivation was significantly achieved at higher pressure treatment of 500 MPa for 10 min. Therefore it was observed that enhancing the treatment time as well as the pressure exposure can augment the high-pressure processing's lethal effect, ascribed to its technological aspects, the composition of food and as well as factors in action in synergy (Ferreira et al., 2016; Possas et al., 2017).

3.4 YOGURT

The exposure of food particles to processing can yield bacterial cell injury as a result of the heat treatment process (Tripathy and Giri, 2014). A fermented probiotic product like yogurt should bear excellent quality properties like textural, gustatory and other sensory properties. In the food industry, in order to retain the textural properties of yogurt, various additives like skimmed milk powder, polysaccharides, caseinates or even whey protein concentrates are incorporated (Lucey, 2002; Leroy and De Vuyst, 2004). It has been explained by researchers that high-pressure processing can retain texture by providing the protein gel with a structurally reinforcing benefit (Trujillo et al., 2002; Anema et al., 2005; Penna et al., 2007; Masson et al., 2011; Tsevdou et al., 2013; Loveday et al., 2013). The fortification of solids in milk can be replaced by applying high-pressure processing at 100 to 400 MPa for a time period of 10 to 15 min at a temperature of 10 to 25° C. This kind of high-pressure product exposure or treatment can significantly improve sensory qualities as well as at the same time retaining the biological properties of yogurt containing live probiotic cells (Tanaka and Hatanaka, 1992; Krompkamp et al., 1995; de Ancos et al., 2000).

Tsevdou et al. (2020) studied the effect of the high-pressure processing of pre-coagulated milk on rheological, sensory and physicochemical properties without causing any sorts of disturbances to probiotic cells' viability. They had chosen microorganisms like *Lactobacillus casei* and *Bifidobacterium bifidum* under various pH conditions like 4.8 and 6.5. The high-pressure processing feasibility was assessed by deriving kinetic models in products like cherry-flavored as well as plain yogurt, which contains probiotic strains like *Lactobacillus casei* and *Bifidobacterium bifidum*.

The exposure of stirred yogurt to high-pressure processing treatment at the rate of 100–300 MPa helps to maintain the desired prescribed level of probiotic counts, i.e. 10^6 CFU/g (FAO/WHO, 2002). The application of high pressure in the last stage of the production process had significantly enhanced the rheological behavior of the yogurt. The plain yogurt sample pressurized at 100–300 MPa displayed both structural as well as textural improvement with respect to coagulum thickness, firmness and uniformity of product. Therefore, from this study, it was concluded that the treating of yogurt with high pressure at 200–300 MPa at the end of the production process can significantly improve the sensory and quality aspects of the final product, ultimately enhancing viscosity and reducing whey separation or syneresis (Tsevdou et al., 2020).

Aryan, a kind of yogurt-based drink, when subjected to a high-pressure processing treatment of 600 MPa for a time period of 5 min, had significantly decreased the microbial cell count of *L. innocua* and *L. monocytogenes* to greater than 5 log units (Evrendilek and Balasubramaniam, 2011). The reduction in count may be because of the baroprotection effect of milk upon cells; this type of effect can counterbalance the microorganism-preventive effect of chokeberry pomace, which further explains the lesser inactivation levels noticed for *L. monocytogenes* in this study.

3.5 CHEESE

Koca et al. (2018) subjected Turkish-based white cheese to high-pressure processing ranging from 50 to 400 MPa for time periods of 5 and 15 min, and after treatment, these samples were brine-ripened for two months. The research was conducted to identify the salt distribution changes in different layers of cheese (internal, external and center) as a result of high-pressure treatment at the above conditions, and the uptake of salt by the cheese block during the ripening process was also analyzed. Their results show that the exposure of pressure-treated and brined white cheese samples doesn't alter the moisture and salt content in various layers of cheese. It seems that an insignificant salt distribution effect was noticed as a result of pressure treatment; even high-pressure exposure did not remarkably bring any salt distribution changes in layers of cheese or a similar effect for salt uptake during the cheese ripening process. The aged cheese after two months' time exhibited a slight enhancement of moisture content for pressure-processed samples of 200 to 400 MPa. The application of high pressure can even change the textural properties of white cheese. Soft-textured product was obtained at a pressure treatment of 200 and 400 MPa for a period of 15 min.

3.6 CONCLUSION

High-pressure processing or high hydrostatic processing can retain the nutritional profile of various dairy products in addition to extending its shelf stability, thereby destroying the microbial load significantly, when compared to traditional thermal heat processing. The industry can adopt such techniques to provide quality-enhanced products to consumers.

REFERENCES

Anema, S.G.; Lauber, S.; Lee, S.K.; Henle, T.; Klostermeyer, H. Rheological properties of acid gels prepared from pressure- and transglutaminase-treated skim milk. *Food Hydrocoll.* 2005, 19(5), 879–887.

Barraquio, V.L. Which milk is fresh? *Int. J. Dairy Process. Res.* 2014, 1, 1–6.

Bogahawaththa, D.; Buckow, R.; Chandrapala, J.; Vasiljevic, T. Comparison between thermal pasteurization and high pressure processing of bovine skim milk in relation to denaturation and immunogenicity of native milk proteins. *Innov. Food Sci. Emerg. Technol.* 2018, 47, 301–308.

Chawla, R.; Patil, G.R.; Singh, A.K. High hydrostatic pressure technology in dairy processing: A review. *J. Food Sci. Technol.* 2011, 48(3), 260–268.

Considine, K.M.; Kelly, A.L.; Fitzgerald, G.F.; Hill, C.; Sleator, R.D. High-pressure processing - Effects on microbial food safety and food quality. *FEMS Microbiol. Lett.* 2008, 281(1), 1–9.

Corrales, M.; Toepfl, S.; Butz, P.; Knorr, D.; Tauscher, B. Extraction of anthocyanins from grape by-products assisted by ultrasonics, high hydrostatic pressure or pulsed electric fields: A comparison. *Innov. Food Sci. Emerg. Technol.* 2008, 9(1), 85–91.

de Ancos, B.; Pilar Cano, M.P.; Gómez, R. Characteristics of stirred low-fat yoghurt as affected by high pressure. *Int. Dairy J.* 2000, 10(1–2), 105–111.

Elamin, W.M.; Endan, J.B.; Yosuf, Y.A.; Shamsudin, R.; Ahmedov, A. High pressure processing technology and equipment evolution: A review. *J. Eng. Sci. Technol. Rev.* 2015, 8(5), 75–83.

Diez-Sánchez, Elena; Martínez, A.; Rodrigo, D.; Quiles, A.; Hernando, I. Optimizing high pressure processing parameters to produce milkshakes using chokeberry pomace. *Foods* 2020, 9(4), 405; doi:10.3390/foods9040405.

Evrendilek, G.A.; Balasubramaniam, V.M. Inactivation of *Listeria monocytogenes* and *Listeria innocua* in yogurt drink applying combination of high pressure processing and mint essential oils. *Food Control* 2011, 22(8), 1435–1441.

Ferreira, M.; Almeida, A.; Delgadillo, I.; Saraiva, J.; Cunha, Â. Susceptibility of Listeria monocytogenes to high pressure processing: A review. *Food Rev. Int.* 2016, 32(4), 377–399.

Food and Agriculture Organization of the United Nations/World Health Organization (FAO/WHO). *Report of a Joint FAO/WHO Working Group on Drafting Guidelines for the Evaluation of Probiotics in Food*; Food and Agriculture Organization of the United Nations/World Health Organization: London, ON, Canada, 2002.

Gao, G.; Ren, P.; Cao, X.; Yan, B.; Liao, X.; Sun, Z.; Wang, Y. Comparing quality changes of cupped strawberry treated by high hydrostatic pressure and thermal processing during storage. *Food Bioprod. Process.* 2016, 100, 221–229.

Hernández-Carrión, M.; Hernando, I.; Quiles, A. High hydrostatic pressure treatment as an alternative to pasteurization to maintain bioactive compound content and texture in red sweet pepper. *Innov. Food Sci. Emerg. Technol.* 2014, 26, 76–85.

Koca, N.; Ramaswamy, R.; Balasubramaniam, W.M.; Harper, W.J. Salt distribution and salt uptake during ripening in Turkish white cheese affected by high pressure processing. *Turkish JAF Sci.Tech.* 2018, 6(4), 433–437.

Krompkamp, J.; Moreira, R.M.; Langeveld, L.P.M.; Van Mil, P.J.J.M. Microorganisms in milk and yoghurt: Selective inactivation by high hydrostatic pressure. In: *Proceedings of the IDF Symposium Heat Treatments and Alternative Methods*, Vienna, Austria, 6–8 September 1995.

Leroy, F.; De Vuyst, L. Lactic acid bacteria as functional starter cultures for the food fermentation industry. *Trends Food Sci. Technol.* 2004, 15(2), 67–78.

Loveday, S.M.; Sarkar, A.; Singh, H. Innovative yoghurts: Novel processing technologies for improving acid milk gel texture. *Trends Food Sci. Technol.* 2013, 33(1), 5–20.

Lucey, J.A. ADSA Foundation Scholar Award. Formation and physical properties of milk protein gels. *J. Dairy Sci.* 2002, 85(2), 281–294.

Masson, L.M.P.; Rosenthal, A.; Calado, V.M.A.; Deliza, R.; Tashima, L. Effect of ultra-high pressure homogenization on viscosity and shear stress of fermented dairy beverage. *LWT Food Sci. Technol.* 2011, 44(2), 495–501.

Patterson, M.F. Microbiology of pressure-treated foods. *J. Appl. Microbiol.* 2005, 98(6), 1400–1409.

Penna, A.L.B.; Subbarao-Gurram, G.V.; Barbosa-Cánovas, G.V. High hydrostatic pressure processing on microstructure of probiotic low-fat yogurt. *Food Res. Int.* 2007, 40(4), 510–519.

Possas, A.; Pérez-Rodríguez, F.; Valero, A.; García-Gimeno, R.M. Modelling the inactivation of *Listeria monocytogenes* by high hydrostatic pressure processing in foods: A review. *Trends Food Sci. Technol.* 2017, 70, 45–55.

Ritz, M.; Tholozan, J.; Federighi, M.; Pilet, M. Physiological damages of Listeria monocytogenes treated by high hydrostatic pressure. *Int. J. Food Microbiol.* 2002, 79(1–2), 47–53.

San Martín, M.F.; Barbosa-Cánovas, G.V.; Swanson, B.G. Food processing by high hydrostatic pressure. *Crit. Rev. Food Sci. Nutr.* 2002, 42(6), 627–645.

Tamime, A.Y. *Milk Processing and Quality Management*; Wiley-Blackwell: West Sussex, UK, 2014; ISBN 9781444301649.

Tan, S.F.; Chin, N.L.; Tee, T.P.; Chooi,S.K. Physico-chemical changes, microbiological properties, and storage shelf life of cow and goat milk from industrial high-pressure processing. *Processes* 2020, 8(697), 1–13. doi:10.3390/pr8060697.

Tanaka, T.; Hatanaka, K. Application of hydrostatic pressure to yoghurt to prevent its after-acidification. *J. Jpn. Soc. Food Sci.* 1992, 39, 173–177.

Tokuşoğlu, Ö. Effect of high hydrostatic pressure processing strategies on retention of antioxidant phenolic bioactives in foods and beverages—A review. *Pol. J. Food Nutr. Sci.* 2016, 66(4), 243–251.

Tripathi, M.K.; Giri, S.K. Probiotic functional foods: Survival of probiotics during processing and storage. *J. Funct. Foods* 2014, 9, 225–241.

Trujillo, A.J.; Capellas, M.; Saldo, J.; Gervilla, R.; Guamis, B. Applications of high-hydrostatic pressure on milk and dairy products: A review. *Innov. Food Sci. Emerg.* 2002, 3(4), 295–307.

Tsevdou, M.S.; Eleftheriou, E.G.; Taoukis, P.S. Transglutaminase treatment of thermally and high pressure processed milk: Effects on the properties and storage stability of set yoghurt. *Innov. Food Sci. Emerg.* 2013, 17, 144–152.

Tsevdou, M.; Ouli-Rousi, M.; Soukoulis, C.; Taoukis, P. Impact of high-pressure process on probiotics: Viability kinetics and evaluation of the quality characteristics of probiotic yoghurt. *Foods* 2020, 9(3), 360.

Vázquez-Gutiérrez, J.L.; Plaza, L.; Hernando, I.; Sánchez-Moreno, C.; Quiles, A.; de Ancos, B.; Cano, M.P. Changes in the structure and antioxidant properties of onions by high pressure treatment. *Food Funct.* 2013, 4(4), 586.

Wang, C.; Huang, H.; Hsu, C.; Yang, B.B. Recent Advances in food processing using high hydrostatic pressure recent advances in food processing using high hydrostatic pressure technology. *Food Sci. Nutr.* 2015, 56, 527–540.

Welti-Chanes, J.; López-Malo, A.; Palou, E.; Bermúdez, D.; Guerrero-Beltrán, J.A.; Barbosa-Cánovas, G.V., 2005 Fundamentals and applications of high pressure processing to foods. In: *Novel Food Processing Technologies*; Barbosa-Cánovas, G.V., Tapia, M.S., Cano, M.P., Martín-Belloso, O., Martínez, A., Eds.; CRC Press: Boca Raton, FL; pp. 164–170.

Chapter 4

Cold Plasma

An Emerging Technology in Milk and Dairy Product Processing

Dharini Manoharan and Mahendran Radhakrishnan

CONTENTS

4.1 Introduction	43
4.2 Types of Cold Plasma	44
4.3 Plasma on Food Materials	44
4.3.1 Cold Plasma for Inactivating Microbial Load	45
4.3.2 Cold Plasma for Modification of Food Components	45
4.4 Milk and Dairy Products	46
4.4.1 Cold Plasma on Milk	47
4.4.2 Cold Plasma on Cheese	59
4.4.3 Plasma on Milk Protein	60
4.4.4 Plasma on Dairy-Based Beverages	61
4.5 Conclusion	62
References	62

4.1 INTRODUCTION

Ionized gas generated at room temperature consists of several reactive species such as ions, radicals, electrons, and electromagnetic radiation (UV rays) known as cold plasma. The possible means of supplying energy to a gas mixture is through the mechanical, electrical, chemical, or radiant source of energy. Plasma possesses reactive species in the ground and excited states with the combination of neutral, negative, and positive charges that give a net neutral charge for plasma (Sharma and Singh 2020). Unlike thermal plasma, in non-thermal plasma, energy is channeled particularly to the electrons rather than energizing the whole gas; this creates a thermal disequilibrium between the electrons and other reactive species (Feizollahi, Misra, and Roopesh 2021). The generation of low-temperature plasma with a high reactivity of species makes it suitable for treating biologic materials that are sensitive to heat (Mir et al. 2019). Further, cold plasma is eco-friendly and non-chemical as it utilizes only ambient air or other gas as an operating substance.

    Cold plasma application on food products is vast, which includes decontamination (Guo, Huang, and Wang 2015), disinfestation (Ratish Ramanan, Sarumathi, and Mahendran 2018), property modification (Zhu 2017), pesticide (Ranjitha Gracy, Gupta, and Mahendran 2019), enzyme activity (Thirumdas, Sarangapani, and Annapure 2015), toxicity (Gavahian and Cullen 2020), and allergen reduction (Ekezie, Sun, and Cheng 2019). It is widely used in almost all food sectors, from wastewater treatment to plasma-assisted toxicity reduction in foods. Dairy is one of the sectors that have been widely researched using non-thermal techniques to maintain/increase the quality of processed milk (Amaral et al. 2017). Cold plasma–associated works have also been conducted on milk and dairy products to increase their safety and standard. Dairy products are primarily packed with beneficial nutrients such as protein, fat, lactose, and other micronutrients. The study of plasma reactive species' interaction with those of the nutrients is of prime significance as plasma is highly reactive with macro- and micromolecules. This chapter offers an outline of the prominence of plasma processing of food materials, its application for maintaining safety, and plasma-assisted modification of products related to the dairy sector.

4.2 TYPES OF COLD PLASMA

Cold plasma, otherwise called non-thermal plasma (NTP), is generated at ambient or near-ambient temperature (Mir et al. 2019). The emergence of plasma generation at atmospheric temperature is a breakthrough that made it easy for application on heat-fragile materials. NTP can be produced by different means, and its three major classifications are based on the sort of plasma discharge, gas, and pressure used for generation. The discharge mechanism for plasma generation includes dielectric barrier discharge (DBD) (produced between electrodes), corona discharge (produced near sharp edges), microwave plasma (remote plasma produced separately and passed to a chamber for treatment), and plasma jet (produced by both DBD or corona discharge methods with the addition of gas flow to produce a jet for direct application) (Sakudo, Misawa, and Yagyu 2019). Plasma can be generated in atmospheric and low-pressure environments with variability in the amount of power level used for generation (a low-pressure environment requires less power for the ionization of gas) (Shashikanthalu, Ramireddy, and Radhakrishnan 2020). Further, plasma generation with different gases (air, helium, oxygen, argon, CO_2, and nitrogen) can also produce differences in plasma reactive species and concentration (Perinban, Orsat, and Raghavan 2019; Zhang et al. 2019). Overall, the three major classifications of plasma clearly explain that plasma can be generated by different means and can be manipulated (types of reactive species required and type of application based on the product) according to the requirement of plasma application.

4.3 PLASMA ON FOOD MATERIALS

Food materials are commercially heat processed to assure safety and increase shelf-life; however, the heat sensitivity of nutrients in food material makes heat processing less favorable. The advent of non-thermal processing (without heat) reduces this problem of nutrient loss and the maintenance of safety. Cold plasma is a non-thermal, non-residual technology that effectively maintains food safety and nutrition (López et al. 2019). In the case of plasma processing of foods, the process parameters along with the properties of food are also crucial in deciding the efficiency of processing. The parameters such as type of food material (solid or liquid), moisture content, dimension (quantity of liquid in case of liquids), porosity, density, and texture play a notable role in plasma interaction with food. The added advantage of using liquid and moisture-rich food is the generation of long shelf-life reactive species through a secondary chemical reaction with liquids (Surowsky, Schlüter, and Knorr 2015; Aparajhitha and Mahendran 2019).

4.3.1 Cold Plasma for Inactivating Microbial Load

Cold plasma is composed of various elements that are distinctly reactive with microorganisms. The primary site for reactive species' activity is the cell wall, its proteins, and its DNA (Bermudez-Aguirre 2019). The reactive species such as free radicals, free electrons, ions, and ionized elements synergistically react with the microbial cell wall, penetrate it, and change the metabolic activity of the microbial cell (Phan et al. 2017). Further, the electric field formed and the ultraviolet radiation generated also react with microorganisms for inactivation. The reactive oxygen and nitrogen species, which are the predominant components of plasma, are the major bactericidal agents of plasma. The reactive species' bombardment of the bacterial cell wall causes chemical modifications (abstraction of hydrogen), erosion, and the rupture of the cell wall (Bourke et al. 2017). The bacterial cell wall is made up of peptidoglycan that consists of bonds such as C–H, C–O, C–N, and C–C; the reactive species dissociates these bonds to disrupt the cell wall. The charged particles tend to accumulate near the cell wall and cause electrostatic disruption, which ultimately leads to electroporation (Schottroff et al. 2018). Apart from physical damage to the microbial cell, the reaction species' penetration through the cell wall favors the oxidization of proteins and lipids and damages the microbial cell. Ultraviolet radiations are mostly responsible for damaging bacterial DNA and photodesorption. The different mechanisms of plasma reactive species that are responsible for microbial inactivation are graphically illustrated in Figure 4.1. Thus, the combined effect of various plasma elements on microbial cells makes the microorganisms hard to recover and produces resistance (Lackmann et al. 2013).

4.3.2 Cold Plasma for Modification of Food Components

Apart from the reduction in microbial load, cold plasma reacts readily with food components such as protein, carbohydrate, and fat to oxidize them; oxidation tends to change the vital properties of the food material. Protein, carbohydrate, fat are essential components in food due to their contribution in providing nutrition, functional properties, and desirable quality attributes. Altering the nature of these macromolecules can enhance the food material's properties (Spicer and Davis 2014). The most beneficial effects of cold plasma treatment on proteins include the inactivation of enzymes, changes in the hydrophobic or hydrophilic nature of the protein, and manipulation of functional, thermal, and cooking properties (Mirmoghtadaie, Shojaee Aliabadi, and Hosseini 2016). Modification of a protein structure might include changes such as side-chain modification, protein backbone modification, fragmentation, crosslinking, and conformation changes (Bjoern Surowsky et al. 2013). Enzymes and allergens are also a form of proteins that can also be altered by cold plasma to rise or decrease the activity (Han, Cheng, and Sun 2019; Tolouie et al. 2018). In the case of carbohydrates, it is necessary to modify the native starch to get improved digestibility and other functional properties; cold plasma–assisted depolymerization (fragmentation), crosslinking, etching, and creation of functional groups on target surfaces can amplify the properties of carbohydrates (Banura et al. 2018). Reactive oxygen and nitrogen species are the main components that are involved in the depolymerization and oxidation of carbohydrates. Though plasma reaction with proteins and carbohydrates is beneficial in improving food properties, its interaction with lipids is not desirable (Gavahian et al. 2018). The reactive species-assisted oxidation in lipids takes place in the double bonds of unsaturated fatty

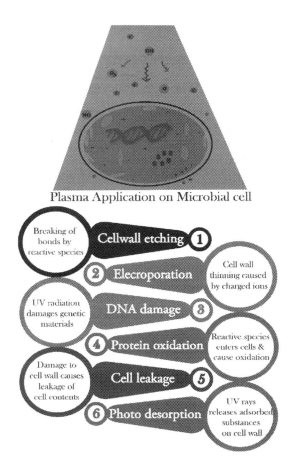

Figure 4.1 Different types of mechanisms involved in microbial reduction by plasma reactive species.

acids. It increases the aldehyde content in food, which in turn produces an unpleasant flavor and reduces the quality (Zhang et al. 2019).

Further, plasma interaction with bioactive compounds/micromolecules such as vitamins, phenolic compounds, minerals, and antioxidants is also gaining importance as plasma tends to enhance the quantity of these micromolecules. However, the increase or decrease of a property or a bioactive compound depends entirely on the process parameter utilized in generating plasma. Different equipment and processing conditions used for treatment change the effect of the plasma on the food component.

4.4 MILK AND DAIRY PRODUCTS

The dairy sector includes various products ranging from milk, skimmed milk, toned and double-toned milk, flavored milk, and dairy products such as butter, ghee, cheese, paneer, dairy cream, ice cream, and milk powders, and protein isolates. Among these dairy products, milk, flavored milk, cheese, yogurt, and ice cream are the most widely studied commodities due to their wide use and nutrient content, which acts as an easy target for microorganisms. In the case of milk (liquid), it has a lesser shelf-life as its nutrition mixture is an essential medium for various microorganisms. Harmful microorganisms act on these liquid foods and spoil them, reducing shelf-life (Vantarakis et al. 2011). Cheese is another important product known for foodborne illness and outbreak worldwide caused by microorganisms such as *Listeria monocytogenes*, *Escherichia coli*, and *Salmonella typhimurium* (Yong et al. 2015b). Apart from the microbial load, maintenance of properties such as emulsification, foaming, and thermal properties is also important to maintain the dairy products' quality.

Though cold plasma effectively decreases the microbial load and maintains the nature of food materials with its interaction with food components, only limited studies have been performed on milk and dairy products. Microbial reduction in milk and dairy products holds the major area compared to quality maintenance. As milk is rich in protein, plasma's properties in modifying the protein properties have numerous potentials in elevating the quality. However, plasma processing parameters must be optimized to sustain the health-beneficial fatty acids in milk, which are highly susceptible to reactive species. The following are the studies that have been conducted on the plasma processing of milk and dairy products. The process parameters used and their effect on microbial load, macro- and micromolecules, physiochemical, functional, and thermal properties are listed in Table 4.1.

4.4.1 Cold Plasma on Milk

As discussed earlier, liquid foods have an added benefit of secondary reactive species origination within liquids, which aids in producing long-lived reactive species. Initially, a study on milk was conducted using a 9 kV corona discharge–type plasma system for reducing the *E. coli* load in 15 ml each of whole milk, semi-skimmed milk, and skimmed milk (Gurol et al. 2012). Milk fat was not observed to hinder the plasma; approximately 56% of the microbial reduction was obtained for 3 min plasma exposure in all milk samples. The *E. coli* log reduction obtained after 20 min of plasma exposure time was 4.15 log CFU/ml. In another study, an encapsulated DBD plasma with 15 kHz, 250 W power was used for reducing the inoculated microbial load (*E. Coli*, *L. monocytogenes*, and *S. Typhimurium*) in 10 ml of milk (Kim et al. 2015). Overall, the maximum microbial reduction obtained was 2.4 log CFU/ml at an exposure time of 10 min with an infinitesimal change in pH and color value (L* and a* value decreased and b* increased). No prominent changes were obtained in the fatty acid composition and the value of thiobarbituric acid reactive substances (TBARS). Korachi et al. (2015) used a 9 kV DBD plasma unit for evaluating the biochemical changes in milk after plasma exposure. The fatty acid profile in milk was increased after 3 and 6 min of plasma exposure time; however, a severe treatment time of 20 min reduced the unsaturated fatty acids.

Using nanosecond-pulse, argon gas corona discharge (pulsed) plasma (Ponraj et al. 2017), the total microbial load of 6.8 logs CFU/ml was inactivated to 1.9 log CFU/ml at 9 kV power level for 2 min exposure time. Further, the milk's shelf life was observed to be stable for six weeks when stored at refrigeration temperature. The amount of hydrogen peroxide present in the sample increased with the increase in plasma exposure time. Another study was done on the microbial reduction in milk on a continuous scale using DBD-type air plasma at low pressure. The plasma produced was applied to the milk flowing under atmospheric pressure (through a quartz tube in between the electrodes) (Manoharan, Stephen, and Radhakrishnan 2020). The system was employed at 2 kV, and the flow rate of milk (3 and 6 mL/min) was adjusted to alter the exposure

TABLE 4.1 COMPILATION OF THE INFLUENCES OF PLASMA TREATMENT ON MILK AND DAIRY PRODUCTS: MICROBIAL CHANGES, PHYSIOCHEMICAL CHANGES, CHANGES IN PROTEINS AND FATS, FUNCTIONAL PROPERTIES, THERMAL PROPERTIES, AND BIOACTIVE COMPOUNDS

S. No.	Commodity	Plasma Parameters Used	Changes	Reference
			Effect of Cold Plasma on Microbial Changes	
1.	Milk	Argon plasma at power 2, 3, and 4 kHz for 2 min exposure time, and 4 kHz for 30 to 120 s	Initial microbial load of 6.8 reduced to 1.9 log CFU/mL at 4 kHz and 120 s treatment	(Ponraj et al. 2017)
2.		Corona discharge, 9 kV, 3–20 min plasma exposure	56% E. coli was reduced for 3 min exposure and at 20 min exposure, 4.1 log reduction was achieved	(Gurol et al. 2012)
3.		250 W and 15 kHz power, DBD plasma unit (encapsulated)	E. Coli, L. monocytogenes, and S. Typhimurium load decreased by approx. 2.4 log CFU/ml at 10 min	(Kim et al. 2015)
4.		Low-pressure DBD continuous type plasma, 2 kV, 57 mA Milk flow rate – 3 and 6 mL/min	E. coli reduction accounted for 1.38 log CFU/mL at 3 mL/min milk flow rate	(Manoharan, Stephen, and Radhakrishnan 2020)
5.	Cheese	Pre-packed condition, DBD, 15 kHz, 100 W peak power and 2 W average power	L. monocytogenes, S. Typhimurium, and E. coli were alleviated by 2.1, 5.8, and 3.2 log CFU/g after 10 min treatment	(Yong et al. 2015b)
6.		DBD, 60 Hz, power 30, 50, and 70 W, 10 min exposure time and 2, 1.5, and 1 cm electrode distance	50 W and 10 min inactivated the E. coli load completely and achieved partial reduction (0.7 log CFU/g at 50 W and 1.6 log CFU/g at 75 W) of Listeria innocua in cheese	(Huang, Chen, and Hsu 2020)
7.		Encapsulated atmospheric pressure DBD chamber, 15 kHz, bipolar with a square wave, 250 W	L. monocytogenes, S. Typhimurium, and E. coli reduced by 2.2, 3.1, and 2.8 log reduction for 15 min exposure. Post-treatment time of 3 min effectively removed viable cells from S. Typhimurium and E. coli followed by 5 min for L. monocytogenes	(Yong et al. 2015a)
8.		Atmospheric plasma jet, 50 kHz, square wave at 3.5 kV, helium and helium + oxygen plasma	1.98 and 1.47 reduction in E. coli was achieved for helium + oxygen and pure helium plasma after 15 min plasma treatment	(Lee et al. 2012)

(Continued)

4.4 Milk and Dairy Products

TABLE 4.1 (CONTINUED) COMPILATION OF THE INFLUENCES OF PLASMA TREATMENT ON MILK AND DAIRY PRODUCTS: MICROBIAL CHANGES, PHYSIOCHEMICAL CHANGES, CHANGES IN PROTEINS AND FATS, FUNCTIONAL PROPERTIES, THERMAL PROPERTIES, AND BIOACTIVE COMPOUNDS

Effect of Cold Plasma on Microbial Changes

S. No.	Commodity	Plasma Parameters Used	Changes	Reference
9.		13.56 MHz radiofrequency, DBD plasma, 75 to 150 W; helium gas at 10 liters per minute flow rate	The *L. monocytogenes* species was reduced to a maximum of 8 log reduction at 150 W and 120 s plasma exposure time	(Song et al. 2009)
10.		Cheese pre-packed with flushed, dry air, DBD, 100 kV, 50 Hz, direct and indirect exposure	Microbial reduction in queso fresco cheese and model cheeses were 3.5 and 1.6 log CFU/g at direct exposure and 2.2 and 0.8 log CFU/g at indirect exposure for 5 min exposure time	(Wan et al. 2019)

Effect of Cold Plasma on the Physicochemical Property of pH

S. No.	Commodity	Plasma Parameters Used	Changes	Reference
1.	Cheese	Cheese pre-packed with flushed, dry air, DBD, 100 kV, 50 Hz, direct and indirect exposure	A slight decrease in pH was obtained in cheese due to the buffering capacity of proteins and fats	(Wan et al. 2019)
2.		Pre-packed condition, DBD, 15 kHz, 100 W peak power and 2 W average power	Significant decrease in pH was noticed after 5 min treatment	(Yong et al. 2015b)
3.	Milk	Low-pressure DBD continuous type plasma, 2 kV, 57 mA Milk flow rate – 3 and 6 mL/min	Infinitesimal reduction in pH was reported after treatment; the conductivity was found to get increased owing to the diffusion of reactive species into milk	(Manoharan, Stephen, and Radhakrishnan 2020)
4.	Whey protein isolate	DBD, 70 kV, exposure time 1, 5, 10, 15, 30, and 60 min Whey protein isolate prepacked with polyethylene terephthalate	A slight alleviation in pH was presented after plasma exposure	(Segat et al. 2015)

(Continued)

TABLE 4.1 (CONTINUED) COMPILATION OF THE INFLUENCES OF PLASMA TREATMENT ON MILK AND DAIRY PRODUCTS: MICROBIAL CHANGES, PHYSIOCHEMICAL CHANGES, CHANGES IN PROTEINS AND FATS, FUNCTIONAL PROPERTIES, THERMAL PROPERTIES, AND BIOACTIVE COMPOUNDS

Effect of Cold Plasma on the Physicochemical Property of pH

S. No.	Commodity	Plasma Parameters Used	Changes	Reference
5.	Casein and whey protein isolates	Spark discharge: 8 kV, 25 kHz Glow discharge: 5 kV, 25 kHz Plasma exposure time: 0 to 30	pH reduced drastically for casein and whey protein solutions from 7 and 7.3 to 4.1 and 4.8, respectively	(Ng et al. 2021)
6.	Sodium caseinate powder	DBD, 7 kV, 10 kHz, exposure time 0 to 10 min	Sodium caseinate suspension showed a gradual reduction in pH with a rise in exposure time	(Jahromi et al. 2020)
7.	Chocolate milk beverage	Low-pressure DBD, 400 W, 50 kHz, nitrogen plasma, at several exposure times and gas flow rates	pH reduced at several processing conditions due to dissolution of reactive species	(Coutinho et al. 2019b)
8.	Guava flavored whey beverage	Low-pressure DBD, 400 W, 50 kHz, nitrogen plasma, at several exposure times and gas flow rates	Plasma-processed beverage increased in pH value compared to the raw sample and pasteurized product. Buffering capacity might be associated with the change in pH	(Silveira et al. 2019b)
9.	Milk	DBD plasma (encapsulated) with 250 W and 15 kHz power	pH reduced slightly after treatment (6.9 to 6.6 at 10 min exposure time)	(Kim et al. 2015)
10.		9 kV corona discharge plasma system, 3 to 20 min exposure	No noticeable changes were observed in pH	(Gurol et al. 2012)
11.		Argon plasma, power at 2, 3, and 4 kHz for 2 min and 4 kHz for 30 to 120 s	pH of milk reduced from 6.76 to 6.67 pH for 4 kHz and 120 s	(Ponraj et al. 2017)
12.	Alkaline phosphatase	DBD, 40, 50 and 60 kV, exposure time 15 s to 5 min	Plasma treatment did not affect the pH of alkaline phosphatase	(Segat et al. 2016)

(Continued)

4.4 Milk and Dairy Products 51

TABLE 4.1 (CONTINUED) COMPILATION OF THE INFLUENCES OF PLASMA TREATMENT ON MILK AND DAIRY PRODUCTS: MICROBIAL CHANGES, PHYSIOCHEMICAL CHANGES, CHANGES IN PROTEINS AND FATS, FUNCTIONAL PROPERTIES, THERMAL PROPERTIES, AND BIOACTIVE COMPOUNDS

Effect of Cold Plasma on the Physiochemical Property of Colour

S. No.	Commodity	Plasma Parameters Used	Changes	Reference
1.	Milk	9 kV corona discharge unit, exposure time from 3 to 20 min	The plasma-treated sample's total color change was 0.25 for 9kV and 9 min exposure time. For 20 min exposure, the change was recorded as 0.52 (slightly noticeable color change)	(Gurol et al. 2012)
2.		DBD plasma (encapsulated) with 250 W and 15 kHz power	Plasma treatment changes the L*, a*, and b* values of milk. L* and b* were increased, and a* was decreased after treatment	(Kim et al. 2015)
3.		Low-pressure DBD continuous type plasma, 2 kV, 57 mA, milk flow rate 3 and 6 mL/min	The overall color change was more than 1.5, which falls under the noticeable color scale. Settlement of fat on the processing tube might be the reason for the decrease in color	(Manoharan, Stephen, and Radhakrishnan 2020)
4.	Cheese	DBD, 60 Hz, power 30, 50, and 70 W, exposure time 10 and 2 min, electrode distance 1.5 and 1 cm	L* value and a* value reduced with a rise in time, whereas the b* increased. The overall color change at 2 and 1 cm electrode distance was 1.5–1.8 and 2.3–2.6	(Huang, Chen, and Hsu 2020)
5.		Pre-packed condition, DBD, 15 kHz, 100 W peak power and 2 W average power	The plasma-treated cheese showed that L* and b* values reduced and a* value increased. The elevation in a* value and fall in L* was attributed to the formation of brown color (non-enzymatic browning)	(Yong et al. 2015b)
6.		Atmospheric plasma jet, 50 kHz, square wave at 3.5 kV, helium and helium + oxygen plasma	L* value decreased, and b* value elevated with an increase in plasma exposure time	(Lee et al. 2012)
7.	Sodium caseinate	DBD, 7 kV, 10 kHz, exposure time 0 to 10 min	Noticeable changes were reported in the b* value of sodium caseinate powder. b* value increased from 7.51 to 8.41 and 8.88 at 5 and 10 min plasma exposure	(Jahromi et al. 2020)

(Continued)

TABLE 4.1 (CONTINUED) COMPILATION OF THE INFLUENCES OF PLASMA TREATMENT ON MILK AND MILK DAIRY PRODUCTS: MICROBIAL CHANGES, PHYSIOCHEMICAL CHANGES, CHANGES IN PROTEINS AND FATS, FUNCTIONAL PROPERTIES, THERMAL PROPERTIES, AND BIOACTIVE COMPOUNDS

Effect of Cold Plasma on the Physiochemical Property of Colour

S. No.	Commodity	Plasma Parameters Used	Changes	Reference
8.	Whey protein isolate	70 kV, DBD, exposure time 1, 5, 10, 15, 30, and 60 min, isolate prepacked with polyethylene terephthalate	b* value of the isolate was observed to raise with the increase in exposure time. Maillard reaction was attributed to the formation of yellowness	(Segat et al. 2015)

Effect of Cold Plasma on Physiochemical Properties: Particle Size, Viscosity, and Consistency

S. No.	Commodity	Plasma Parameters Used	Changes	Reference
1.	Chocolate beverage	Low-pressure DBD, 400 W, 50 kHz, nitrogen plasma, at several exposure times and gas flow rates	Cold plasma treatment produced chocolate beverages with reduced particle size and increased volume diameters. 20 mL/min for 15 min treatment resulted in particles with the maximum surface area. Reduced particles size increases the creaminess (consistency) of the beverage	(Coutinho et al. 2019a)
2.	Guava based whey beverage	Nitrogen gas plasma, low-pressure DBD, 80 kV, flow rate 10, 20, and 30 mL/min, exposure time 5, 10, and 15 min	Severe processing conditions reduced the particle size and increased the number of particles in the beverage. Milder conditions resulted in a product with bigger fat globules, which were similar to the pasteurized product. In the case of viscosity, intermediate processing conditions reduced the viscosity and consistency, whereas, at severe process conditions, the viscosity and consistency were higher	(Silveira et al. 2019b)
3.	Milk	Low-pressure DBD continuous type plasma, 2 kV, 57 mA, milk flow rate 3 and 6 mL/min	The viscosity of milk shows a significant difference from the raw milk. The reduction in the particle size during plasma application might be associated with the reduction in viscosity as it causes a lubricating effect in milk	(Manoharan, Stephen, and Radhakrishnan 2020)

(Continued)

TABLE 4.1 (CONTINUED) COMPILATION OF THE INFLUENCES OF PLASMA TREATMENT ON MILK AND DAIRY PRODUCTS: MICROBIAL CHANGES, PHYSIOCHEMICAL CHANGES, CHANGES IN PROTEINS AND FATS, FUNCTIONAL PROPERTIES, THERMAL PROPERTIES, AND BIOACTIVE COMPOUNDS

Effect of Cold Plasma on Protein: Structure and Allergenicity

S. No.	Commodity	Plasma Parameters Used	Changes	Reference
1.	Milk casein and whey proteins	25 kHz plasma unit Spark discharge: 8 kV Glow discharge: 5 kV Plasma exposure time: 0 to 30	SDS-PAGE disclosed that the casein and α-lactalbumin decreased after plasma exposure with a formation of a higher molecular weight band (denotes aggregation or crosslinking). β-lactoglobulin bands were found to increase after treatment. The secondary structure of FTIR analysis showed that the protein structure was unfolded after treatment. β-sheets reduced with a rise in α-helix and β-turns. Amino acid profile was observed to be alleviated after treatment. ELISA test revealed that the allergenicity of casein and α-lactalbumin reduced, and the allergenicity of β-lactoglobulin raised after plasma treatment	(Ng et al. 2021)
2.	Whey protein and casein	Argon gas plasma, 13.56 MHz RF power, maximum exposure time 15 min	No significant modifications in ELISA and SDS-PAGE were obtained after treatment	(Tammineedi et al. 2013)
3.	Sodium caseinate	DBD, 7 kV, 10 kHz, exposure time 0 to 10 min	FTIR analysis of sodium caseinate disclosed that the random coils were reduced with an increase in the exposure time. At 2.5 and 5 min exposure, the β-sheets were observed to rise. However, it decreased at 10 min plasma exposure	(Jahromi et al. 2020)

(Continued)

TABLE 4.1 (CONTINUED) COMPILATION OF THE INFLUENCES OF PLASMA TREATMENT ON MILK AND DAIRY PRODUCTS: MICROBIAL CHANGES, PHYSIOCHEMICAL CHANGES, CHANGES IN PROTEINS AND FATS, FUNCTIONAL PROPERTIES, THERMAL PROPERTIES, AND BIOACTIVE COMPOUNDS

Effect of Cold Plasma on Protein: Structure and Allergenicity

S. No.	Commodity	Plasma Parameters Used	Changes	Reference
4.	Whey protein isolate	DBD, 70 kV, exposure time 1, 5, 10, 15, 30, and 60 min, whey protein isolate prepacked with polyethylene terephthalate	The carbonyl content in the plasma processed isolate was more than the untreated isolate. The sulfhydryl group was observed to be reduced to half at 10 min of plasma exposure, which implies disulphide bonds during plasma application. The hydrophobic nature of the isolate was also found to increase significantly with respect to plasma exposure time; it reveals that there was protein unfolding in the course of plasma treatment. Significant increase in particle size and the polydispersity index values were also reported after 30 and 60 min	(Segat et al. 2015)
5.	Alkaline phosphatase	40, 50, and 60 kV, DBD unit, 15 s to 5 min	The secondary structure of the protein was modified at treatments 40 and 60 kV for 0, 120, and 300 s. α-helical structure was identified to decrease (unfolding) with a rise in voltage and time	(Segat et al. 2016)

Effect of Cold Plasma on Fat and Its Composition

S. No.	Commodity	Plasma Parameters Used	Changes	Reference
1.	Cheese	Pre-packed condition, DBD, 15 kHz, 100 W peak power and 2 W average power	The TBARS value, which denotes fat oxidation, was increased significantly and gradually with plasma processing time. TBARS increased from 0.13 in the raw sample to 0.16 and 0.18 in 5 and 10 min of plasma exposure time	(Yong et al. 2015b)
2.		DBD, 60 Hz, power 30, 50, and 70 W, exposure time 10 and 2 min, 1.5 and 1 cm electrode distance	The TBARS value of plasma-treated cheese was found to be not notably different from the untreated cheese	(Huang, Chen, and Hsu 2020)

(Continued)

TABLE 4.1 (CONTINUED) COMPILATION OF THE INFLUENCES OF PLASMA TREATMENT ON MILK AND DAIRY PRODUCTS: MICROBIAL CHANGES, PHYSIOCHEMICAL CHANGES, CHANGES IN PROTEINS AND FATS, FUNCTIONAL PROPERTIES, THERMAL PROPERTIES, AND BIOACTIVE COMPOUNDS

Effect of Cold Plasma on Fat and Its Composition

S. No.	Commodity	Plasma Parameters Used	Changes	Reference
3.	Chocolate milk beverage	Low-pressure DBD, 400 W, 50 kHz, nitrogen plasma, at several exposure times and gas flow rates	The mild and severe process conditions of plasma decreased the unsaturated fatty acids (oxidation) and increased the saturated fatty acids. Parallelly, the intermediate processing condition improved the fatty acid profile	(Countinho et al. 2019b)
4.	Guava based whey beverage	Low-pressure DBD, nitrogen gas plasma, 80 kV, flow rate 10, 20, and 30 mL/min, 5, 10, and 15 min exposure time	Unsaturated fatty acids were found to decrease after plasma treatment. The abstraction of hydrogen atoms from unsaturated fatty acids might be the reason for more reduction	(Silveira et al. 2019a)
5.	Milk	DBD plasma (encapsulated) with 250 W and 15 kHz power	No significant changes were obtained in the fatty acid profile of plasma-treated milk. Further, the TBARS value was also noted to be similar to the untreated milk sample	(Kim et al. 2015)
6.		9 kV corona discharge, exposure time 3, 6, 9, 12, 15, and 20 min	The fatty acid profile of milk increased after 3 and 6 min exposure time; however, severe treatment time of 20 min reduced the unsaturated fatty acids. Oleic acid in milk increased from 23.9 % to 24.2 and 24.7 at 3 and 6 min plasma exposure	(Korachi et al. 2015)

Effect of Cold Plasma on Functional Properties

S. No.	Commodity	Plasma Parameters Used	Changes	Reference
1.	Whey protein isolate	DBD, 70 kV, exposure time 1, 5, 10, 15, 30, and 60 min, whey protein isolate prepacked with polyethylene terephthalate	Foaming and emulsion properties increased gradually till 15 min of treatment. However, 30 and 60 min of treatment decreased the properties. The change in protein structure and unfolding was attributed to the changes in the functional properties	(Segat et al. 2015)

(Continued)

TABLE 4.1 (CONTINUED) COMPILATION OF THE INFLUENCES OF PLASMA TREATMENT ON MILK AND DAIRY PRODUCTS: MICROBIAL CHANGES, PHYSIOCHEMICAL CHANGES, CHANGES IN PROTEINS AND FATS, FUNCTIONAL PROPERTIES, THERMAL PROPERTIES, AND BIOACTIVE COMPOUNDS

Effect of Cold Plasma on Thermal Properties

S. No.	Commodity	Plasma Parameters Used	Changes	Reference
1.	Sodium caseinate	DBD, 7 kV, 10 kHz, exposure time 0 to 10 min	It was reported that the glass transition temperature (T_g – mobility of protein) increase gradually in 2.5 and 5 min treatment, and it reduced at 10 min. The denaturation temperature (T_d) increased at 2.5 and 5 min and no T_d peak was found at 10 min treatment. The elevation in sheets in protein structure at 2.5 and 5 min and its decrease at 10 min must be associated with the change in T_g and T_d	(Jahromi et al. 2020)
2.	Chocolate milk beverage	Low-pressure DBD, 400 W, 50 kHz, nitrogen plasma, at several exposure times and gas flow rates	Higher enthalpy, lower melting point, and lower bound water were secured after plasma treatment. Higher enthalpy denotes the increase in denaturation following cold plasma treatment, and the creation of aggregates was attributed to the inbound reduction of water	(Coutinho et al. 2019a)
3.	Guava based whey beverage	Low-pressure DBD, 400 W, 50 kHz, nitrogen plasma, at several exposure times and gas flow rates	A significant increase in melting temperature was reported after treatment. The enthalpy values were not found significantly different from the pasteurized sample. The bound water decreased with the increase in exposure time and gas flow rate	(Silveira et al. 2019b)

(Continued)

4.4 Milk and Dairy Products

TABLE 4.1 (CONTINUED) COMPILATION OF THE INFLUENCES OF PLASMA TREATMENT ON MILK AND DAIRY PRODUCTS: MICROBIAL CHANGES, PHYSIOCHEMICAL CHANGES, CHANGES IN PROTEINS AND FATS, FUNCTIONAL PROPERTIES, THERMAL PROPERTIES, AND BIOACTIVE COMPOUNDS

Effect of Cold Plasma on Bioactive Components and Nutrients

S. No.	Commodity	Plasma Parameters Used	Changes	Reference
1.	Chocolate milk beverage	Low-pressure DBD, 400 W, 50 kHz, nitrogen plasma, at several exposure times and gas flow rates	The total phenolic content of the plasma-treated sample was reduced in comparison with the pasteurized product. The modification might be ascribed to the presence of ozone in plasma, as phenols are very sensitive to ozone. The antioxidant property of plasma-treated beverages also decreased after processing. The angiotensin-converting enzyme (ACE) inhibitory activity rose with an increase in processing time and flow rate used for plasma treatment. A maximum of 13.75% of inhibition took place at 30 ml/min flow rate and 15 min exposure time	(Coutinho et al. 2019b)
2.	Milk	Low-pressure DBD continuous type plasma, 2 kV, 57 mA, milk flow rate 3 and 6 mL/min	Nutrients such as protein, lactose, and calcium were not notably different from raw milk. Fat content was observed to reduce drastically from 3.5 in raw milk to 2 in plasma-processed milk. The deposition of fat on the process tube surface due to the continuous processing of milk could be the reason for the reduction	(Manoharan, Stephen, and Radhakrishnan 2020)
3.	Guava based whey beverage	Low-pressure DBD, nitrogen gas plasma, 80 kV, flow rate 10, 20, and 30 mL/min, exposure time 5, 10, and 15 min	Cold plasma processing of guava-based whey beverage increased the quantity of carotenoids, vitamin C, phenols, ACE inhibitory activity, and antioxidant capacity. The rise in phenolic compound was ascribed with the disintegration of the cell membrane, which increases the release of phenols from cell membranes. Vitamin C density was more at the milder and intermediate process conditions. An increase in reactive species at severe process conditions was due to the reduction of vitamin C	(Silveira et al. 2019a)

time. A maximum of 95% microbial reduction was acquired at a 3 mL/min flow rate, which continuously processed 180 mL of milk per hour. The physicochemical variables such as pH, conductivity, color, titratable acidity, and viscosity were examined, which revealed that significant changes were obtained only for the color and viscosity of milk. Further, the elucidation of plasma on milk nutrients such as protein, calcium, lactose, and fat disclosed significant changes in color and viscosity due to the decrease in fat percentage. The practicable reason reported for the change in fat content was its deposition on the plasma processing tubes during processing.

The studies on plasma application for milk disclose that cold plasma is an efficient technology in reducing milk microbial load. Plasma processing was also observed to have the potential to sustain the quality of the milk. However, most of the studies' quantity of milk used for plasma processing was smaller compared to quantities used for other non-thermal processing methods. This could affect the processing condition and the efficacy of processing when similar equipment must be scaled up to process milk. Further studies must be experimented with to increase the amount of milk processed to obtain a clear picture of the plasma's interaction with microbial reduction. Various plasma-assisted changes that are imparted on milk and dairy products are pictorially represented in Figure 4.2.

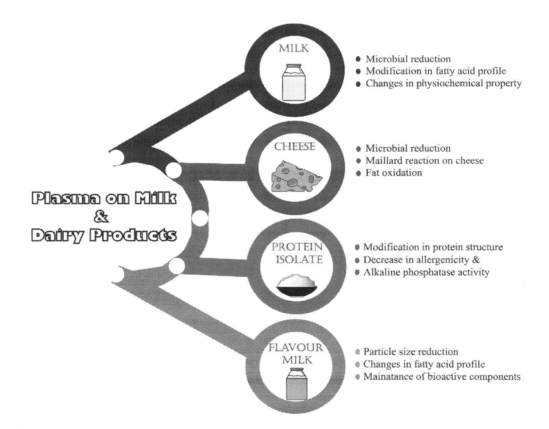

Figure 4.2 Pictorial representation of the major influences of plasma on milk and dairy products.

4.4.2 Cold Plasma on Cheese

Cheese is a major dairy product that is processed by the coagulation of milk protein. It is produced worldwide using different mammal milk with various flavors and textures. Cold plasma processing of cheese was mainly done to alleviate the microbial load and assess lipid oxidation. A study on cheese slices was conducted with atmospheric pressure plasma to reduce three selected strains of *Listeria* spp. at different power levels and exposure times (Song et al. 2009). The microbial load in this study decreased with a rise in power level and exposure time. Maximum microbial reduction of 8 log CFU/g was obtained at 150 W power level for 120 s exposure time with a D-value of 17.27 s for 90% microbial inactivation.

Similarly, another study was conducted on cheese slices to assess the impact of different carrier gas concentrations in a plasma jet on microbial reduction (Lee et al. 2012). Helium as a pure gas and a mixture of helium and oxygen was utilized at a flow rate of 4 mL per minute, and the plasma was produced at a 50 kHz, square wave for 3.5 kV. It was described that the microbial log reduction in the plasma treatment with the admixture carrier gas was more compared to the pure helium gas plasma. The maximum *E. coli* reduction was achieved at 15 min exposure time with 1.98 and 1.47 log CFU/g for helium and oxygen and pure helium plasma.

An atmospheric pressure encapsulated DBD chamber made of parallelly arranged copper electrodes on a plastic container (250 W, 15 kHz, bipolar with a square wave) was employed to lessen the microbial load on cheese slices and agar plates (Yong et al. 2015a). The log reduction of microorganisms in agar was higher than cheese, which might be due to the differences such as texture, moisture, and plasma adsorption between agar and cheese. *L. monocytogenes*, *S. Typhimurium*, and *Escherichia coli* load were reduced by 2.2, 3.1, and 2.8 log reduction for 15 min. Apart from the treatment, the study was also conducted for estimating the effect of post-treatment time (post-retention time after plasma exposure) on microbial reduction. It was identified that the increase in post-treatment duration (5 min) increased the microbial load, which was reported to be due to the phenomena of maintenance of ozone concentration similar to the case of intermediate plasma *on* and *off* conditions. During the process, switching off the system preserved the ozone formed when the power was on.

Cheddar is a kind of cheese that is one of the most popular varieties consumed worldwide. Cheddar cheese was treated using a DBD plasma unit (100 W maximum power level, 15 kHz, and 2 W average power) for lowering the microbial load (Yong et al. 2015b); bacterial strains such as *L. monocytogenes*, *S. Typhimurium*, and *E. coli* were reduced by 2.1, 5.8, and 3.2 log CFU/g after 10 min of plasma exposure. Physiochemical parameter assessment revealed that the pH and the color value, lightness (L* value) decreased, whereas TBARS and yellowness (b* value) of milk increased with plasma exposure time. However, the overall color change and sensory values did not show any significant difference. Another variety called queso fresco cheese, model cheese, and tryptic soy agar were studied to reduce the *Listeria innocua* load (inoculated) using high-voltage atmospheric plasma (Wan et al. 2019). The cheese samples, which were pre-packed with flushed, dry air, were exposed to plasma (DBD, 50 Hz, 100 kV) under both direct (between DBD electrodes) and indirect exposure (away from electrodes). The microbial reduction in queso fresco cheese and model cheese were 3.5 and 1.6 log CFU/g at direct exposure and 2.2 and 0.8 log CFU/g at indirect exposure for 5 min exposure time, whereas there was 5 log reduction in tryptic soy agar at direct exposure. This variation in microbial reduction was correlated to the surface roughness and microstructure of the samples; the rise in surface roughness reduces the microbial reduction in samples. Further, the decrease in pH was more in tryptic soy agar compared to other samples, which corresponds to the percentage of protein and fat content in the samples.

Similarly, a DBD-type atmospheric plasma was employed to reduce the microbial load on cheese slices; the rise in power level and the exposure time were found to elevate the microbial reduction in cheese (Huang, Chen, and Hsu 2020). Complete inactivation of *E. coli* and partial inactivation (0.7 log CFU/g at 50 W and 1.6 log CFU/g at 75 W) of *L. innocua* was noted at 50 W, 2 cm electrode distance, and 10 min treatment. Further, a minor change was observed on hardness and color after 10 min treatment time. However, reducing the electrode distance was observed to reduce the processing time in deactivating the microorganisms; 99.99% microbial inactivation was obtained at 1 min for 1 cm and 3 min for 1.5 cm electrode distance.

Among the six studies of cold plasma treatment on cheese, the peak microbial reduction was obtained at a maximum power study (8 log CFU/g). The experiments on the inclusion of helium and oxygen gas for the generation of plasma disclosed the importance of oxygen's availability in the process gas. The experiments on the assessment of the type of plasma exposure and electrode distance revealed the importance of direct exposure and reduced electrode distance in microbial reduction. Further, the b* value of cheese was found to increase in two studies, which indicates the possibility of plasma-assisted non-enzymatic browning.

4.4.3 Plasma on Milk Protein

Milk proteins and their concentrates can be used for value addition in cheese making, nutritional products, infant dairy formula, and other dairy beverages. Despite the nutritional benefits of these proteins, they have low solubility and functional properties. Thus, to boost the nature of dairy proteins, they are subjected to different processing techniques. As cold plasma is very reactive with protein, the application can enhance the properties of dairy proteins. The influence of plasma reactive species on protein is highly complex and involves changes in the protein structure, which ultimately modify proteins' properties. Milk contains two major proteins, casein and whey proteins (α-lactalbumin and β-lactoglobulin).

Whey protein isolate was studied for evaluating the changes imparted on its functionality after cold plasma treatment (Segat et al. 2015). The study on the carbonyls and sulfhydryl groups on plasma-treated whey protein isolates disclosed that the amino acids in the isolates were modified, which was denoted by the rise in the carbonyl groups and also in the depletion of the sulfhydryl groups. Sodium caseinates are proteins that are extracted after the removal of calcium phosphate from the casein micellar protein. Sodium caseinates are subjected to different treatments for improving their properties; cold plasma was also used to assess its impact on the functional, thermal, and structural properties of caseinates (Jahromi et al. 2020). An atmospheric pressure DBD unit with a 7 kV power supply and 10 kHz was used for treating sodium caseinate power for 0, 2.5, 5, and 10 min exposure time. It was noticed that the exposure time of 5 min was optimal in obtaining caseinate film with an even structure and increased thermal and mechanical properties. The FTIR analysis uncovered that the plasma exposure time increased the β sheets and decreased the helical structure, and the DSC analysis proved higher glass transition after treatment.

Allergens in food materials are generally proteins, which act as antigens and bind with the IgE antibodies available in the human body and release harmful substances such as histamine. The major proteins in milk, such as lactoglobulin, lactalbumin, and casein, are food allergens that cause allergies in certain sensitized individuals. The influence of plasma (spark and glow discharge) on reducing this allergenicity was studied using SDS-PAGE, ELISA, FTIR, and MS/MS (Ng et al. 2021). SDS-PAGE bands displayed that the intensities of casein and α-lactalbumin alleviated after plasma treatment, whereas no changes were observed in β-lactoglobulin bands.

Following the SDS-PAGE result, the ELISA test disclosed that the allergenicity decreased in α-lactalbumin and casein, whereas it increased in β-lactoglobulin after treating with plasma. The FTIR analysis revealed that the protein's secondary structure was altered after treatment; it was responsible for the modification in allergenicity of the milk proteins. Alkaline phosphatase is a major enzyme present in milk. The influence of plasma in alleviating the activity of alkaline phosphatase was studied at various power levels (40, 50, and 60 kV) and different time durations (15 s to 5 min) (Segat et al. 2016). The elimination of enzyme activity was achieved within a few seconds of treatment. It was distinguished in the circular dichroism that the structure of the enzyme changed with the plasma treatment; the α-helical structure was raised, and β structures decreased with changes in power and exposure time.

The studies of cold plasma treatment on dairy proteins reveal that the plasma reactive species aids in unfolding the protein structure. Thus, plasma elevates the properties of proteins and reduces the allergenicity and enzyme activity of dairy proteins, which are the major concern for dairy-allergen-sensitized people.

4.4.4 Plasma on Dairy-Based Beverages

Dairy-based beverages can be prepared by introducing flavors to milk or with the addition of dairy proteins to other juices. The property enhancement of guava-flavored whey beverage and chocolate milk was experimented on using cold plasma. Cold plasma on guava-based whey beverage was evaluated for changes in vitamin C, DPPH, phenolic compounds, fatty acid profile, carotenoids, and volatile compounds (Silveira et al. 2019a). At milder conditions of reduced airflow rate and plasma exposure time, vitamin C content, antioxidant activity, and volatile components increased. However, carotenoid content was reduced, and the fatty acid profile was found unfavorable. Higher processing conditions resulted in reduced vitamin C content and volatile compounds, whereas carotenoid content increased, and the enzyme activity was also found to be improved after treatment. In another study, plasma treated guava-flavored whey-based beverage was investigated in terms of changes in parameters such as pH, viscosity, consistency, and flow behavior (Silveira et al. 2019b). In mild treatment conditions, the pH increased (dissolution of reactive species), and the viscosity and consistency decreased due to partial disintegration of particles, whereas in higher treatment conditions, viscosity and consistency increased, which could be associated with the increased reduction in particle size and smaller fragments.

Similar to the guava-based whey beverage, chocolate milk beverage was also studied for investigating the impact of cold plasma on pH, viscosity, particle size distribution, consistency, and thermal properties of milk (Coutinho et al. 2019a) and DPPD, ACE inhibitory activity, fatty acid profile, phenolic compounds, and volatile compounds of milk (Coutinho et al. 2019b). In milder conditions, the viscosity, particle size, and consistency of the beverage increased, which indicates the aggregation in the chocolate beverage, whereas ACE inhibitory activity, phenolic compounds, and favorable fatty acid profile decreased at both severe and milder treatment conditions. However, at an intermediate flow rate and process time, the particle size reduced compared to the milder and severe process conditions. The product obtained was similar to thermal pasteurized milk in terms of microstructure, consistency, thermal properties, and particle size. The bioactive compounds like ACE inhibitory activity, nutritional quality, fatty acid profile, antioxidant activity, and volatile compounds were improved compared to pasteurized milk.

In both the cases of guava and the chocolate-based dairy beverage, plasma treatment at intermediate processing condition was noticed to be better than milder and severe processing conditions at upholding the quality attributes of the beverages. The works on microbial reduction, changes in properties, and nutrient compounds in dairy products using plasma changed with respect to the type of plasma processing equipment, the power supplied, operating gas, gas flow rate, exposure time, and the properties of food materials. Though optimization of these parameters was done in all these studies, they differ from one another based on plasma exposure efficiency and its interaction with dairy products. In the case of milk whey protein and casein isolates, the allergenicity was observed to reduce in one study and not in another. A detailed study on these topics is essential for obtaining knowledge on the parameters that alter the activity of the protein isolates.

4.5 CONCLUSION

Non-thermal processing of food materials is gaining importance as these technologies can be applied for sensitive food material, which is highly susceptible to heat processing. Cold plasma is one of the non-thermal technologies that has been broadly studied to sustain the safety, nutritional and functional quality of food. Cold plasma was found effective in inactivating the microbial load and improving the quality of milk and other dairy products, which are packed with nutritious components. The administration of plasma on milk and dairy products is crucial due to the existence of nutrients in dairy products that are highly reactive with plasma. Large quantities of unsaturated fatty acid make it an easy target for reactive species-assisted oxidation. However, the lack of intense research work on the influence of plasma on milk and dairy products compared to other food materials makes it hard to get closure on the apt changes that cold plasma imparts on dairy products.

REFERENCES

Amaral, Gabriela V., Eric Keven Silva, Rodrigo N. Cavalcanti, Leandro P. Cappato, Jonas T. Guimaraes, Verônica O. Alvarenga, Erick A. Esmerino, Jéssica B. Portela, Anderson S. Sant' Ana, Mônica Q. Freitas, Márcia C. Silva, Renata S.L. Raices, M. Angela A. Meireles and Adriano G. Cruz. 2017. "Dairy Processing Using Supercritical Carbon Dioxide Technology: Theoretical Fundamentals, Quality and Safety Aspects." *Trends in Food Science and Technology* 64. Elsevier Ltd: 94–101. doi:10.1016/j.tifs.2017.04.004.

Aparajhitha, S., and R. Mahendran. 2019. "Effect of Plasma Bubbling on Free Radical Production and Its Subsequent Effect on the Microbial and Physicochemical Properties of Coconut Neera." *Innovative Food Science and Emerging Technologies* 58. Elsevier Ltd. doi:10.1016/j.ifset.2019.102230. 102230.

Banura, Sidhant, Rohit Thirumdas, Amritpal Kaur, R. R. Deshmukh, and U. S. Annapure. 2018. "Modification of Starch Using Low Pressure Radio Frequency Air Plasma." *LWT - Food Science and Technology* 89. Elsevier Ltd: 719–24. doi:10.1016/j.lwt.2017.11.056.

Bermudez-Aguirre, Daniela. 2019. "Advances in the Inactivation of Microorganisms and Viruses in Food and Model Systems Using Cold Plasma." In: *Advances in Cold Plasma Applications for Food Safety and Preservation* Elsevier Inc: 49–91. doi:10.1016/B978-0-12-814921-8.00002-5.

Bourke, Paula, Dana Ziuzina, Lu Han, P. J. Cullen, and B. F. Gilmore. 2017. "Microbiological Interactions with Cold Plasma." *Journal of Applied Microbiology* 123(2). Wiley Online Library: 308–24.

Coutinho, Nathalia M., Marcello R. Silveira, Tatiana C. Pimentel, Monica Q. Freitas, Jeremias Moraes, Leonardo M. Fernandes, Marcia C. Silva, Renata S.L. Raices, C. Senaka Ranadheera, Fábio O. Borges, Roberto P.C. Neto, Maria Inês B. Tavares, Fabiano A.N. Fernandes, Filomena Nazzaro, Sueli

Rodrigues, and Adriano G. Cruz. 2019a. "Chocolate Milk Drink Processed by Cold Plasma Technology: Physical Characteristics, Thermal Behavior and Microstructure." *LWT* 102: 324–9. doi:10.1016/j.lwt.2018.12.055.

Coutinho, Nathalia M., Marcello R. Silveira, Leonardo M. Fernandes, Jeremias Moraes, Tatiana C. Pimentel, Monica Q. Freitas, Marcia C. Silva, Renata S.L. Raices, C. Senaka Ranadheera, C. Senaka Ranadheera, Fábio O. Borges, Roberto P.C. Neto, Maria Inês B. Tavares, Fabiano A.N. Fernandes, Thatyane V. Fonteles, Filomena Nazzaro, Sueli Rodrigues, and Adriano G. Cruz. 2019b. "Processing Chocolate Milk Drink by Low-Pressure Cold Plasma Technology." *Food Chemistry* 278. Elsevier Ltd: 276–83. doi:10.1016/j.foodchem.2018.11.061.

Ekezie, Flora Glad Chizoba, Da Wen Sun, and Jun Hu Cheng. 2019. "Altering the IgE Binding Capacity of King Prawn (Litopenaeus Vannamei) Tropomyosin through Conformational Changes Induced by Cold Argon-Plasma Jet.. " *Food Chemistry* 300(July). Elsevier: 125143. doi:10.1016/j.foodchem.2019.125143.

Feizollahi, E., N. N. Misra, and M. S. Roopesh. 2021. "Factors Influencing the Antimicrobial Efficacy of Dielectric Barrier Discharge (DBD) Atmospheric Cold Plasma (ACP) in Food Processing Applications." *Critical Reviews in Food Science and Nutrition* 61(4): 666–89. doi:10.1080/10408398.2020.1743967.

Gavahian, M., Y. Chu, A. Mousavi Khaneghah, F. J. Barba, and N. N. Misra. 2018. "A Critical Analysis of the Cold Plasma Induced Lipid Oxidation in Foods." *Trends in Food Science and Technology*. doi:10.1016/j.tifs.2018.04.009.

Gavahian, Mohsen, and P. J. Cullen. 2020. "Cold Plasma as an Emerging Technique for Mycotoxin-Free Food: Efficacy, Mechanisms, and Trends." *Food Reviews International* 36(2). Taylor & Francis: 193–214. doi:10.1080/87559129.2019.1630638.

Guo, Jian, Kang Huang, and Jianping Wang. 2015. "Bactericidal Effect of Various Non-Thermal Plasma Agents and the in Fl Uence of Experimental Conditions in Microbial Inactivation : A Review." *Food Control* 50. Elsevier Ltd: 482–90. doi:10.1016/j.foodcont.2014.09.037.

Gurol, C., F. Y. Ekinci, N. Aslan, and M. Korachi. 2012. "Low Temperature Plasma for Decontamination of E. coli in Milk." *International Journal of Food Microbiology* 157(1). Elsevier B.V.: 1–5. doi:10.1016/j.ijfoodmicro.2012.02.016.

Han, Y. X., J. H. Cheng, and D. W. Sun. 2019. "Changes in Activity, Structure and Morphology of Horseradish Peroxidase Induced by Cold Plasma." *Food Chemistry*. Elsevier Ltd: 125240. doi:10.1016/j.foodchem.2019.125240.

Huang, Yi Ming, Chung Kai Chen, and Chuan Liang Hsu. 2020. "Non-Thermal Atmospheric Gas Plasma for Decontamination of Sliced Cheese and Changes in Quality." *Food Science and Technology International* 26(8): 715–26. doi:10.1177/1082013220925931.

Jahromi, Mastaneh, Mehrdad Niakousari, Mohammad Taghi Golmakani, Fatemeh Ajalloueian, and Mohammadreza Khalesi. 2020. "Effect of Dielectric Barrier Discharge Atmospheric Cold Plasma Treatment on Structural, Thermal and Techno-Functional Characteristics of Sodium Caseinate." *Innovative Food Science and Emerging Technologies* 66. Elsevier Ltd. doi:10.1016/j.ifset.2020.102542. 102542.

Kim, Hyun-joo, Hae Yong, Sanghoo Park, Kijung Kim, Wonho Choe, and Cheorun Jo. 2015. "Microbial Safety and Quality Attributes of Milk Following Treatment with Atmospheric Pressure Encapsulated Dielectric Barrier Discharge Plasma." *Food Control* 47. Elsevier Ltd: 451–56. doi:10.1016/j.foodcont.2014.07.053.

Korachi, May, Fatma Ozen, Necdet Aslan, Lucia Vannini, Maria Elisabetta Guerzoni, Davide Gottardi, and Fatma Yesim Ekinci. 2015. "Biochemical Changes to Milk Following Treatment by a Novel, Cold Atmospheric Plasma System." *International Dairy Journal* 42. Elsevier Ltd: 64–9. doi:10.1016/j.idairyj.2014.10.006.

Lackmann, Jan Wilm, Simon Schneider, Eugen Edengeiser, Fabian Jarzina, Steffen Brinckmann, Elena Steinborn, Martina Havenith, Jan Benedikt, and Julia E. Bandow. 2013. "Photons and Particles Emitted from Cold Atmospheric-Pressure Plasma Inactivate Bacteria and Biomolecules Independently and Synergistically." *Journal of the Royal Society Interface* 10 (89). doi:10.1098/rsif.2013.0591.

Lee, H., S. Jung, H. Jung, S. Park, W. Choe, J. Ham, and C. Jo. 2012. "Evaluation of a Dielectric Barrier Discharge Plasma System for Inactivating Pathogens on Cheese Slices." *Journal of Animal Science and Technology* 54(3): 191–98. doi:10.5187/jast.2012.54.3.191.

López, Mercedes, Tamara Calvo, Miguel Prieto, Rodolfo Múgica-Vidal, Ignacio Muro-Fraguas, Fernando Alba-Elías, and Avelino Alvarez-Ordóñez. 2019. "A Review on Non-Thermal Atmospheric Plasma for Food Preservation: Mode of Action, Determinants of Effectiveness, and Applications." *Frontiers in Microbiology*. doi:10.3389/fmicb.2019.00622.

Manoharan, Dharini, Jaspin Stephen, and Mahendran Radhakrishnan. 2020. "Study on Low-Pressure Plasma System for Continuous Decontamination of Milk and Its Quality Evaluation." *Journal of Food Processing and Preservation*, e 15138. doi:10.1111/jfpp.15138.

Mir, S. A., M. W. Siddiqui, B. N. Dar, M. A. Shah, M. H. Wani, S. Roohinejad, G. A. Annor, K. Mallikarjunan, C. F. Chin, and A. Ali. 2019. "Promising Applications of Cold Plasma for Microbial Safety, Chemical Decontamination and Quality Enhancement in Fruits." *Journal of Applied Microbiology*. doi:10.1111/jam.14541.

Mirmoghtadaie, Leila, Saeedeh Shojaee Aliabadi, and Seyede Marzieh Hosseini. 2016. "Recent Approaches in Physical Modification of Protein Functionality." *Food Chemistry*. doi:10.1016/j.foodchem.2015.12.067.

Ng, Sing Wei, Peng Lu, Aleksandra Rulikowska, Daniela Boehm, Graham O'Neill, and Paula Bourke. 2021. "The Effect of Atmospheric Cold Plasma Treatment on the Antigenic Properties of Bovine Milk Casein and Whey Proteins." *Food Chemistry*. Elsevier Ltd: 128283. doi:10.1016/j.foodchem.2020.128283.

Paatre Shashikanthalu, Sharanyakanth, Lokeswari Ramireddy, and Mahendran Radhakrishnan. 2020. "Journal of Applied Research on Medicinal and Aromatic Plants Stimulation of the Germination and Seedling Growth of Cuminum Cyminum L. Seeds by Cold Plasma." *Journal of Applied Research on Medicinal and Aromatic Plants* (February). Elsevier. doi:10.1016/j.jarmap.2020.100259. 100259.

Perinban, S., V. Orsat, and V. Raghavan. 2019. "Non-Thermal Plasma–Liquid Interactions in Food Processing: A Review." *Comprehensive Reviews in Food Science and Food Safety* 18(6): 1985–2008. doi:10.1111/1541-4337.12503.

Phan, Khanh Thi Kim, Huan Tai Phan, Charles S. Brennan, and Yuthana Phimolsiripol. 2017. "Non-Thermal Plasma for Pesticide and Microbial Elimination on Fruits and Vegetables: An Overview." *International Journal of Food Science and Technology* 52(10): 2127–37. doi:10.1111/ijfs.13509.

Ponraj, Sri B., Julie A. Sharp, Jagat R. Kanwar, Andrew J. Sinclair, Ladge Kviz, Kevin R. Nicholas, and Xiujuan J. Dai. 2017. "Argon Gas Plasma to Decontaminate and Extend Shelf Life of Milk." *Plasma Processes and Polymers* 14(11): 1–8. doi:10.1002/ppap.201600242.

Ranjitha Gracy, T. K., V. Gupta, and R. Mahendran. 2019. "Influence of Low-Pressure Nonthermal Dielectric Barrier Discharge Plasma on Chlorpyrifos Reduction in Tomatoes." *Journal of Food Process Engineering*: 1–16. doi:10.1111/jfpe.13242.

Ratish Ramanan, K., R. Sarumathi, and R. Mahendran. 2018. "Influence of Cold Plasma on Mortality Rate of Different Life Stages of Tribolium castaneum on Refined Wheat Flour." *Journal of Stored Products Research* 77. Elsevier Ltd: 126–34. doi:10.1016/j.jspr.2018.04.006.

Sakudo, Akikazu, Tatsuya Misawa, and Yoshihito Yagyu. 2019. "Equipment Design for Cold Plasma Disinfection of Food Products." In: *Advances in Cold Plasma Applications for Food Safety and Preservation*: 289–307. Elsevier Inc. doi:10.1016/B978-0-12-814921-8.00010-4.

Schottroff, Felix, Antje Fröhling, Marija Zunabovic-Pichler, Anna Krottenthaler, Oliver Schlüter, and Henry Jäger. 2018. "Sublethal Injury and Viable but Non-Culturable (VBNC) State in Microorganisms during Preservation of Food and Biological Materials by Non-Thermal Processes." *Frontiers in Microbiology* 9: 1–19. doi:10.3389/fmicb.2018.02773.

Segat, Annalisa, N. N. Misra, P. J. Cullen, and N. Innocente. 2015. "Atmospheric Pressure Cold Plasma (ACP) Treatment of Whey Protein Isolate Model Solution." *Innovative Food Science and Emerging Technologies*. doi:10.1016/j.ifset.2015.03.014.

Segat, Annalisa, N. N. Misra, P. J. Cullen, and N. Innocente. 2016. "Effect of Atmospheric Pressure Cold Plasma (ACP) on Activity and Structure of Alkaline Phosphatase." *Food and Bioproducts Processing* 98. Institution of Chemical Engineers: 181–88. doi:10.1016/j.fbp.2016.01.010.

Sharma, Shruti, and Rakesh k. Singh. 2020. "Cold Plasma Treatment of Dairy Proteins in Relation to Functionality Enhancement." *Trends in Food Science and Technology* 102(May). Elsevier: 30–6. doi:10.1016/j.tifs.2020.05.013.

Silveira, Marcello R., Nathalia M. Coutinho, Erick A. Esmerino, Jeremias Moraes, Leonardo M. Fernandes, Tatiana C. Pimentel, Monica Q. Freitas, Márcia C. Silva, Renata S.L. Raices, C. Senaka Ranadheera, Fábio O. Borges, Roberto P.C. Neto, Maria Inês B. Tavares, Fabiano A.N. Fernandes, Thatyane V.

Fonteles, Filomena Nazzaro, Sueli Rodrigues, and Adriano G. Cruz. 2019a. "Guava-Flavored Whey Beverage Processed by Cold Plasma Technology: Bioactive Compounds, Fatty Acid Profile and Volatile Compounds." *Food Chemistry* 279: 120–7. doi:10.1016/j.foodchem.2018.11.128.

Silveira, Marcello R., Nathalia M. Coutinho, Ramon S. Rocha, Jeremias Moraes, Erick A. Esmerino, Tatiana C. Pimentel, Monica Q. Freitas, Márcia C. Silva, Renata S.L. Raices, C. Senaka Ranadheera, Fábio O. Borges, Thatyane V. Fonteles, Roberto P.C. Neto, Maria Inês B. Tavares, Fabiano A.N. Fernandes, Sueli Rodrigues, and Adriano G. Cruz. 2019b. "Guava Flavored Whey-Beverage Processed by Cold Plasma: Physical Characteristics, Thermal Behavior and Microstructure." *Food Research International* 119. Elsevier Ltd: 564–70. doi:10.1016/j.foodres.2018.10.033.

Song, Hyun Pa, Binna Kim, Jun Ho Choe, Samooel Jung, Se Youn Moon, Wonho Choe, and Cheorun Jo. 2009. "Evaluation of Atmospheric Pressure Plasma to Improve the Safety of Sliced Cheese and Ham Inoculated by 3-Strain Cocktail Listeria monocytogenes." *Food Microbiology* 26(4). Elsevier Ltd: 432–36. doi:10.1016/j.fm.2009.02.010.

Spicer, Christopher D., and Benjamin G. Davis. 2014. "Selective Chemical Protein Modification." *Nature Communications* 5. Nature Publishing Group: 1–14. doi:10.1038/ncomms5740.

Surowsky, B., A. Fischer, O. Schlueter, and D. Knorr. 2013. "Cold Plasma Effects on Enzyme Activity in a Model Food System." *Innovative Food Science and Emerging Technologies* 19. Elsevier Ltd: 146–52. doi:10.1016/j.ifset.2013.04.002.

Surowsky, B., Oliver Schlüter, and Dietrich Knorr. 2015. "Interactions of Non-Thermal Atmospheric Pressure Plasma with Solid and Liquid Food Systems: A Review." *Food Engineering Reviews* 7(2): 82–108. doi:10.1007/s12393-014-9088-5.

Tammineedi, Chatrapati V.R.K., Ruplal Choudhary, Gabriela C. Perez-Alvarado, and Dennis G. Watson 2013. "Determining the Effect of UV, High Intensity Ultrasound and Nonthermal Atmospheric Plasma Treatments on Reducing the Allergenicity of α-Casein and Whey Proteins." *LWT - Food Science and Technology*. doi:10.1016/j.lwt.2013.05.020.

Thirumdas, R., C. Sarangapani, and U. S. Annapure. 2015. "Cold Plasma: A Novel Non-Thermal Technology for Food Processing." *Food Biophysics* 10(1): 1–11. doi:10.1007/s11483-014-9382-z.

Tolouie, Haniye, Mohammad Amin Mohammadifar, Hamid Ghomi, and Maryam Hashemi. 2018. "Cold Atmospheric Plasma Manipulation of Proteins in Food Systems." *Critical Reviews in Food Science and Nutrition* 58(15): 2583–97. doi:10.1080/10408398.2017.1335689.

Vantarakis, A., M. Affifi, P. Kokkinos, M. Tsibouxi, and M. Papapetropoulou. 2011. "Occurrence of Microorganisms of Public Health and Spoilage Significance in Fruit Juices Sold in Retail Markets in Greece." *Anaerobe* 17(6): 288–91. doi:10.1016/j.anaerobe.2011.04.005.

Wan, Zifan, S. K. Pankaj, Curtis Mosher, and Kevin M. Keener. 2019. "Effect of High Voltage Atmospheric Cold Plasma on Inactivation of Listeria innocua on Queso Fresco Cheese, Cheese Model and Tryptic Soy Agar." *LWT* 102. Elsevier Ltd: 268–75. doi:10.1016/j.lwt.2018.11.096.

Yong, Hae In, Hyun Joo Kim, Sanghoo Park, Amali U. Alahakoon, Kijung Kim, Wonho Choe, and Cheorun Jo. 2015a. "Evaluation of Pathogen Inactivation on Sliced Cheese Induced by Encapsulated Atmospheric Pressure Dielectric Barrier Discharge Plasma." *Food Microbiology* 46. Elsevier Ltd: 46–50. doi:10.1016/j.fm.2014.07.010.

Yong, Hae In, Hyun Joo Kim, Sanghoo Park, Kijung Kim, Wonho Choe, Suk Jae Yoo, and Cheorun Jo. 2015b. "Pathogen Inactivation and Quality Changes in Sliced Cheddar Cheese Treated Using Flexible Thin-Layer Dielectric Barrier Discharge Plasma." *Food Research International* 69. Elsevier B.V.: 57–63. doi:10.1016/j.foodres.2014.12.008.

Zhang, Kexin, Camila A. Perussello, Vladimir Milosavljević, P. J. Cullen, Da Wen Sun, and Brijesh K. Tiwari. 2019. "Diagnostics of Plasma Reactive Species and Induced Chemistry of Plasma Treated Foods." *Critical Reviews in Food Science and Nutrition* 59(5). Taylor & Francis: 812–25. doi:10.1080/10408398. 2018.1564731.

Zhu, Fan. 2017. "Plasma Modification of Starch." *Food Chemistry* 232(April): 476–86. doi:10.1016/j.foodchem.2017.04.024.

Chapter 5

UV Pasteurization Technology Approaches for Market Milk Processing

Nazia Nissar, Sadaf Rafiq, Rabia Latif, M. Yaseen Sofi, Taibah Bashir and Sheikh Mansoor

CONTENTS

5.1 Introduction	67
5.2 Principle of UV Light Technology	68
5.2.1 Basic Principle	70
5.3 Sources of UV Light Technology	70
5.3.1 Mercury Lamps	71
5.3.2 Excimer Lamps	71
5.3.3 Pulsed UV Lamps	71
5.3.4 Light-Emitting Diodes (LEDs)	72
5.3.5 UV Light Devices	72
5.4 Factors Affecting UV Light Technology in Market Milk	72
5.5 Factors Affecting the Efficacy of UV Light in the Food Industry	73
5.5.1 UV Light Sources	74
5.5.2 UV Light Devices	74
5.5.3 Reactors	74
5.6 The Applications of UV Light in the Dairy Industry	75
5.6.1 Disinfection of Air in the Production Area	75
5.6.2 Disinfection of Water Used During Processing	75
5.6.3 Surface Applications of Packaging Materials and Equipment	75
5.6.3.1 Packaging Materials	75
5.6.3.2 Food Contact Surfaces	75
5.7 Efficacy of UV on Market Milk	75
References	77

5.1 INTRODUCTION

Food handling and protection methods are endlessly being improved to fulfill current consumer demands for safe and beneficial foods. Higher income, development, demographic shifts, expanded conveyance, and consumer judgments regarding quality and safety are changing universal food consumption. Contemporary consumers demand tasty, healthier, natural and fresh-like foods, generated in a biologically friendly manner with endurable procedures and small carbon footprints. As a consequence, in the past two decades non-thermal technologies have obtained rising attention due to their capability for inactivating spoilage and pathogenic microorganisms (Koutchma et al., 2009). Ultraviolet (UV) pasteurization, which is a non-thermal technology, has recently drawn a lot of awareness to the control of foodborne pathogens and spoilage organisms for food safety and shelf-life extension. UV irradiation is considered one of the effective means of disinfection that eliminates the requirement of heat to get rid of microorganisms (Sastry et al., 2008). UV irradiation of milk was first used in the mid-1900s for vitamin D enrichment. The potency of UV light treatment has been studied in recent years and more and more research has also been carried out to analyze the potential applications of UV light as a non-thermal alternative to the thermal processing of milk. On the other hand, due to the inveterate achievement and convenience of thermal processing, potential processing alternatives for milk are still confined. The use of UV light must not only be considered for microbial inactivation but also for the advancement of novel dairy products. UV-treated pasteurized cow's milk was authorized as a novel food in the market by the European Commission. It is documented that the handling of pasteurized milk with UV radiation results in an increase in the vitamin D3 (cholecalciferol) concentrations by conversion of 7-dehydrocholesterol to vitamin D3. In 2012, the dairy apex forfeited recommendations to place pasteurized milk that was treated with UV light on the market as a novel food. This UV procedure of pasteurized milk aspires to broaden its shelf-life from the current 12 days to 21 days and to enhance vitamin D3 concentrations. There is a profitable opening for adopting ultraviolet processing in a small or large-scale food and dairy processing industry. With the approval of the FDA, numerous new applications of UV processing are being assessed and verified by the dairy and food industries in the United States of America. With the potential for offering superior organoleptic qualities of food products at a lower initial investment and operating costs, the authors foresee a great achievement for the adoption of UV processing technology by the food processing industry (Choudhary and Bandla, 2012). In 2013, the pertinent authorities in Ireland sent the commission their initial review documents, which concluded that pasteurized milk treated with UV light meets standards for endorsement as a novel food. In this chapter, the UV pasteurization of milk is illustrated in terms of its principles, mechanisms, and available UV light citations and reactors. Then, the effects of UV light on the inactivation of microorganisms and transformations in the chemical and nutritional aspects of market milk are conferred.

5.2 PRINCIPLE OF UV LIGHT TECHNOLOGY

Thermal processing is the most common method for the destruction of harmful and contagious microorganisms present in food products, which ensures the safety of food products and enhances their shelf-life. Nowadays, in the food industry, the utilization of non-thermal technology (ultraviolet light) as a substitute for thermal processing has been developed. A non-thermal technology called ultraviolet (UV) light has shown increasing effectiveness in improving the safety of food, which nowadays is subject to a lot of awareness. UV light is one of the most important and rising non-thermal technologies of food processing that can be used in producing a wide range of food products with extended shelf-life, protection from any harmful microorganisms and highly nourished quality of food. On the electromagnetic spectrum, the wavelengths of UV light are ranged from 100 to 400 nm. According to the wavelengths, UV light is grouped into four classes: UV-A (315–400 nm) which is responsible for normal tanning on the skin of human beings; UV-B (280–315 nm) which causes skin cancer and can lead to the burning of the skin;

UV-C (200–280 nm) which is the decontamination range that efficiently deactivates microorganisms like bacteria and viruses; and the fourth class is vacuum UV (100–200 nm) which can be transmitted only in vacuum and approximately absorbed by all types of materials. At the earth's surface, among the different classes of UV light, UV-C light is feeble as it is blocked by the ozone layer of the atmosphere. But UV-C rays are considered to be an effective method for deactivating harmful and contagious microorganisms present in dairy and milk products. These rays cause DNA damage due to the formation of lesions in DNA and which results in the destruction of cellular enzyme activity and cytoplasm membrane integrity. The process of UV-C rays' effectiveness mainly depends on the product type (chemical composition, thickness, turbidity, dullness and coarseness), process parameters (source of UV light, exposure, dose and wavelength), microorganism's characteristics (class, strain, initial count, growth phase and recovery conditions) and equipment (conformation and geometry). The highly effective germicidal effect of UV-C light has been reported to be on microorganisms such as bacteria, viruses, protozoa, fungi and algae (Bintsis et al., 2000; Shin et al., 2016). The UV light of wavelength 253.7 nm energy shows the highest amalgamation of DNA and the UV-C radiation (250–260 nm) range has the maximum inhibitory effect on pathogenic and spoilage microorganisms (Bintis et al., 2000).

The principle of UV-C light is that the microorganisms (bio-molecules) show photochemical reactions which result in the deactivation of the cell membrane, cellular activities and thus inhibit the growth of microorganisms due to germicidal effect. This inhibitory effect of UV-C light on pathogenic and spoilage microorganisms occurs due to cross-linking in the same DNA strand among the bases of adjacent pyrimidine dimers (Miller et al., 1999). This germicidal effect ultimately leads to inhibition of transcription and nucleic acid replication, which results in clonogenic death (Bolton and Linden, 2003; Gomez-Lopez et al., 2007). Sometimes, the injured DNA can be repaired by the processes of light-reactivation or dark-reactivation depending on the pathogenic microorganisms involved. But the repair cannot be possible at high UV doses, due to the wider DNA damage.

There are two modes, namely continuous mode and pulse mode, by which UV light can be applied. In the continuous mode, UV light is released as constant energy in a monochromatic or polychromatic wavelength, while in the pulsed mode, the light is released in very short intervals or period pulses after the electrical energy is stored over a small period of time in a capacitor. This electrical energy leads to the ionization of inert gas which is present in a lamp from which it is transferred, releasing a wide range of light spectrum wavelengths in the region of ultraviolet (UV) to near-infrared. Compared to sunlight, pulsed light has 20,000 times more intensity (Dunn et al., 1995). It was addressed that the pulse has a rate of 1 to 20 pulses per second and a width of 300 ns to 1 ms. Therefore, high-megawatt pulsed light energy is produced that is comparable to the total continuous mode of the UV light system. The energy of pulsed UV light is more effective. So, the treatment of pulsed light is a speedier method of deactivating pathogenic and spoilage microorganisms compared to the continuous mode of UV light.

The technology of pulsed light that efficiently deactivates pathogenic and spoilage microorganisms like bacteria, yeasts, molds and even viruses is different for solid and liquid food products. The food products that absorb pulsed light result in the decomposition of light intensity in an exponential mode as the pulsed light has been absorbed according to the depth of food and absorption coefficient of food (Koutchma et al., 2009). It has been observed that due to the effect of high-intensity pulses, pulsed light shows a number of shocking effects on the bacterial cell wall. The effect of pulsed light on liquid samples is more demanding as spoilage and pathogenic microorganisms reside in the full volume of liquids. There are various factors like the distance of the sample from the lamp, exposure, turbidity, optical and physicochemical properties of the sample as well as sample depth that are responsible for the efficient utilization of pulsed light (Ortega-Rivas, 2012; Kasahara et al., 2015). Nowadays, pulsed UV light has gained a lot of attention for microbial deactivation in different food products and also the sterilization of commercial packaging materials with no release of toxic materials (FDA, 2000). The surface of food products or packaging materials, including in-package sterilization if the packaging material can allow UV light to penetrate easily, can be efficiently done by pulsed ultraviolet (UV) light (Butz and Tauscher, 2002).

5.2.1 Basic Principle

Ultraviolet light is a kind of electromagnetic radiation that is responsible for glowing black-light posters, summer tans and sunburns. However, living tissues get damaged due to too much exposure to UV radiation. The natural key source of electromagnetic radiation is the sunlight, which is emitted at various wavelengths and frequencies in the form of waves or particles. The large range of wavelengths is known as the electromagnetic spectrum (EM). The spectrum is normally distributed into seven areas, namely radio waves, microwaves, infrared (IR), visible light, ultraviolet (UV), X-rays and gamma rays, on the basis of decreasing wavelength and increasing energy and frequency. UV radiation is a non-ionizing source of invisible light or any electromagnetic spectrum (EM) light comprising a wavelength range from 100 nm (visible light) to 400 nm (X-rays) (Koutchma et al., 2009; NASA, 2019). UV is generally categorized into four main subtypes, that is, long-wave UV-A or near UV (315–400 nm), responsible for the tanning of skin, UV-B or middle UV (280–315 nm), causing skin burns and ultimately skin cancer, and short-wave UV-C or far UV (200–280 nm), also named the germicidal range because it greatly inactivates pathogenic organisms. Some radiations are also mentioned as vacuum or extreme-UV (100–200 nm). Delorme et al. (2020) has reported that these wavelengths propagate only in a vacuum as they are blocked by air.

5.3 SOURCES OF UV LIGHT TECHNOLOGY

To improve the efficiency of microbial inactivation, it is important to choose the appropriate UV source to increase the penetration of UV light. A number of artificial sources for UV output, namely tanning booths, black lights, curing lamps, germicidal lamps, mercury vapor lamps, halogen lights, high-intensity discharge lamps, fluorescent and incandescent sources and some types of lasers, have been identified by the Health Physics Society. Due to their ability to destroy microorganisms, germicidal lamps are designed to emit UV-C radiation (FDA, 2019). However, mercury lamps (LPM, LPHO-A and MPM), excimer (EL), pulsed (PL) lamps and light-emitting diodes (LED) are commercially available UV sources (Koutchma, 2009). The LPM and excimer lamps are monochromatic sources, while MPM and PL emissions are polychromatic sources. LPM and MPM lamps are the primary sources of UV light for the treatment of fluid foods, beverages and drinks, including water processing.

Lamp manufacturers take into account the following characteristics to compare the electrical and germicidal performance of UV sources:

- Input total power (W): depends on voltage and electric current.
- UV-C lamp efficiency: calculated on the basis of the ratio of measured UV output wattage in the spectral range of interest to the total wattage input to the lamp.

- Irradiance: the quantity of incident flux on a predefined region of the surface. It is most often represented in mW cm^2 or W m^2.
- UV radiance at a given distance: 10 cm or 20 cm for the lamp(s) base.
- Lamp lifetime: when the light irradiance emitted is greater than 80–85% of the initial output value, then the total number of hours of UV source operation is the lamp lifetime.

5.3.1 Mercury Lamps

Mercury vapor UV lamp sources have been successfully used in water treatment for almost 50 years now and are considered to be a reliable source of other disinfection therapies that help with their performance, low cost and quality. There are three different types of mercury UV lamps used: low-pressure (LPM); high-output low-pressure (LPHO); and medium-pressure (MPM) (Kowalski, 2009). When the lamps act as bases to generate UV-C light, these concepts are based on the vapor pressure and ionization of mercury. In food applications, LPM lamps are typically used (Koutchma et al., 2009). LPM lamps have 254 nm of continuous monochromatic light, while MPM lamps emit between 200 and 300 nm of germicidal polychromatic light (Koutchma, 2009). LPM lamps run at a nominal total gas pressure of 102 to 103 Pa, which is equal to the mercury vapor pressure at a temperature of 40° C. At 253.7 nm (85% of total intensity) and 185 nm, the output spectrum of LPM is concentrated at resonance lines. In terms of germicidal effect, the wavelength of 253.7 nm is most efficient since photons are most absorbed by microorganisms' DNA at this particular wavelength (Guerrero-Beltrán and Barbosa-Cánovas, 2004; Lopez-Malo and Palou, 2004). At a total gas pressure of 104 to 106 Pa, the MPM lamps operate. The coolest possible temperature of the MPM is around 400° C relative to the LPM lamps, but it goes up to 600–800° C. The MPM radiation spectrum spans wavelengths from approximately 250 nm to nearly 600 nm. Since MPM lamps have a heavy UV radiation flux that results in a high penetration depth, they are otherwise not considered useful for targeted germicidal therapy. LPHO amalgam lamps containing mercury amalgam have recently been developed and integrated into disinfection applications that are a major development for economic UV-C generation (Schalk et al., 2006), but LPM and MPM are the key sources of UV disinfection care.

5.3.2 Excimer Lamps

These lamps, which can emit pulsed light at 248 nm, are another UV light source. By using different gases such as He, Ne, Ar, Kr and Xe in excimer lamps, it is possible to emit light at the desired wavelength. Even at very low surface temperatures, the excimer lamps can work (Gomez-Lopez et al., 2007).

5.3.3 Pulsed UV Lamps

This is an improved and enhanced version of UV-C light, and due to high intensity, wider range, instant start and durable packaging with no mercury in the lamp, its effectiveness is potentially greater than other sources. Pulsed UV light, as it includes ultraviolet lamps, is an application that uses devices that emit ultraviolet light at high power at regular intervals (Koutchma et al., 2012; Koca et al., 2018). It is applied in the range of 200–1,100 nm in a very short time (1 µs–0.1 s) (Gomez-Lopez et al., 2007). The cumulative effect of photochemical, photothermal and photophysical conditions exists in this technology and microorganisms are ineffective (Gomez Lopez et al., 2012). The emission is equivalent to solar light in wavelength composition. Due to the delivery of high-intensity UV, the UV pulsed devices can penetrate opaque fluids much better than mercury lamps and provide improved treatment speeds. The pulsed light UV technology also uses xenon lamps (Kowalski, 2009). These lamps emit flashes within a short period of time and have a wide range of radiation (180–1,100 nm).

5.3.4 Light-Emitting Diodes (LEDs)

In recent years, UV light-emitting diodes (LEDs) have been produced with many advantages, such as low cost, energy efficiency, long life, easy pollution control and no mercury waste generation. The commercial UV LED has a wavelength range of 240 to 400 nm. An LED is a semiconductor device that emits photon-generating light when carriers with different polarities (electrons and holes) combine or through electroluminescence (Koutchma et al., 2012, 2019). The photon's wavelength depends on the energy gap that the carriers resolve in order to mix (Koutchma, 2019; Schubert, 2006). The UV LED device, which works between 210 nm and 365 nm, is an example of aluminum nitride (AlN), gallium nitride (GaN) and intermediate alloys, which are recent developments in semiconductor materials. This breakthrough was crucial for the further incorporation of UV light technology in applications for food safety. A broad range of UV LEDs are currently available in wavelength ranges from 210 to 400 nm, including several wavelengths in the UV-C germicidal zone (Koutchma, 2019). It is possible that many applications that use mercury lamps today will be carried out by UV LEDs in the near future. Overall, due to its small scale, low power draw and its ability to emit a large range of discrete UV wavelengths, UV LEDs have a substantial advantage over other UV light sources.

5.3.5 UV Light Devices

For the application of UV light to liquids, reactors are used. Inside, the UV reactor contains UV lamps and each UV lamp is placed in a separate protective quartz tube to avoid direct liquid interaction. UV rays emitted from lamps are exposed to the liquid flowing through the UV reactor. Thus, the microorganisms become ineffective in the liquid. The physical, chemical and microbiological properties of the liquid to be disinfected and the volume of the liquid moving through are the most important parameters in the selection of UV reactors. In this context, according to the nature of the fluid and the target microorganism, the UV light dose should be calculated. In addition, parameters such as sediment and liquid turbidity should be extracted by responsive filters to improve the efficiency of disinfection (Ha et al., 2016; Koca et al., 2018). Due to the variations in the location and residence times of microorganisms in some regions of the irradiated area, the flow pattern of liquid in the UV reactor also has a major effect on the total UV dose (Koutchma, 2009). Using turbulent flow in continuous flow UV reactors, microorganism inactivation increases (Koutchma et al., 2004; Franz et al., 2009).

5.4 FACTORS AFFECTING UV LIGHT TECHNOLOGY IN MARKET MILK

Dairy products create good growth conditions for a variety of microorganisms because they are rich in many kinds of nutrients including carbohydrates (especially lactose), lipids, proteins, essential amino acids, enzymes, vitamins and minerals (Haug et al., 2007). Therefore, producing safe dairy products is more challenging compared to producing many other foods. Thermal processing is the most common decontamination method to ensure food safety and to prolong shelf-life by eliminating spoilage and pathogenic microorganisms and enzymes. In recent years, the use of non-thermal technologies is increasing as an alternative to thermal processing in the food industry. Ultraviolet (UV) light, which is a non-thermal technology, has recently attracted a lot of attention with regard to the improvement of food safety. Compared to thermal processing, this promising technology can provide consumers with minimally processed, microbiologically safe and fresh-like products with minor effects on the nutritional and sensory properties of the product. On the other hand, this technology must not replace hygiene, good manufacturing or agricultural practice (Woldemariam and Emire, 2019).

UV light application can also be introduced as an alternative to the use of chemicals in the food industry. Besides, the use of UV light does not generate chemical residues. Additionally, it offers some technological advantages especially in developing countries in a small-scale production due to its low maintenance cost, low installation cost and low operational cost with minimal energy use. The operation and cleaning of the treatment are quite easy. In spite of its many advantages, its low penetration power restricts the area of use in the food industry. Furthermore, its inactivation efficiency may be reduced or prevented because of the physical features of food. At high doses, it can create negative effects on quality and some vitamins. In order to obtain effective results, applications should be made considering these situations. The use of UV light must not only be considered for microbial inactivation but also for the development of novel dairy products. The UV-treated pasteurized cow's milk was authorized as a novel food in the market by the European Commission. It is reported that the treatment of pasteurized milk with UV radiation results in an increase in the vitamin D3 (cholecalciferol) concentrations by conversion of 7-dehydrocholesterol to vitamin D3. Contamination of dairy products with microorganisms may occur at several stages of production, originating from a variety of sources during production (Koutchma, 2019). Although heat treatment is applied for the inactivation of foodborne pathogens, dairy products, especially cheese, can be contaminated with undesirable microorganisms. After the pasteurization process, handling of the curd, equipment, processing lines, packaging or storage rooms can result in cross-contamination with a variety of microorganisms. Even if good manufacturing practices are applied, surface applications of antimicrobial agents before packaging are commonly used to prevent spoilage and extend storage life for some dairy products. Instead of chemical preservatives, an additional solution is needed to control the growth of microorganisms just before or after the packaging of dairy products (Koutchma, 2009). Surface application of UV light after production can offer an attractive alternative method to eliminate or control the growth of post-processing contamination. Other promising uses of UV light are the disinfection of air and water used in a dairy plant, and surface decontamination of food contact surfaces and packaging materials. A lot of research is mainly focused on the application of UV light to reduce microorganisms in milk, and relatively little research focuses on the decontamination of the surfaces of solid dairy products. There is a lack of information about the relation of quality and safety of dairy products (Koutchma, 2019).

5.5 FACTORS AFFECTING THE EFFICACY OF UV LIGHT IN THE FOOD INDUSTRY

The UV light efficacy depends on several factors related to UV equipment, UV sources, operating and measuring conditions, target microorganisms and material or food to be exposed in the food industry (Gomez-Lopez, 2007; Koutchma, 2019) which are summarized as:

- UV light source and UV dose.
- UV sensitivity of microorganisms.

- The composition of the target.
- Physical properties of the target (turbidity, opaque, color, etc.).
- Surface properties of the target (roughness, dirt, etc.).

5.5.1 UV Light Sources

Choosing the right UV source can increase the efficiency of microbial inactivation by increasing UV light penetration (Delorme et al., 2020). The first and natural source of UV light is the sun. The sun emits radiation across a wide range of wavelengths. Other UV light sources are lamps. Many alternative UV light sources have been developed, such as low-pressure mercury (LPM), medium-pressure mercury (MPM), low-pressure high-output mercury lamp-amalgam type, mercury-free amalgam lamps, pulsed-light (PL) and excimer lamps. LPM lamps are commonly used in food applications. Mercury lamps have been the source of radiation in most UV-based disinfection systems. The low- and medium-pressure mercury UV lamp sources are reliable sources for disinfection applications which are beneficial due to their performance and low cost. They are based on the vapor pressure of mercury while the lamps are operating. LPM lamps are designed to deliver a continuous monochromatic light at 254 nm. MPM lamps emit germicidal polychromatic light between 200 and 300 nm. A breakthrough for economic UV-C generation is the discovery of low-pressure amalgam lamps. The mercury emissions from lamps to the environment have encouraged the investigation of mercury-free lamps. Xenon lamps are used in pulsed light UV technology. These lamps emit flashes in a short period of time. They have a broad spectrum of radiation between 180 and 1,100 nm. Another UV light source is excimer lamps, which can emit pulsed light at 248 nm. It is possible to emit light in the desired wavelength by using various gases such as He, Ne, Ar, Kr and Xe in the excimer lamps. The excimer lamps can be operated even at very low surface temperatures (Delorme et al., 2020).

5.5.2 UV Light Devices

UV light applications are carried out with different equipment for solids or liquids: UV reactor designs for liquids according to flow types and UV cabinet designs for solids. It is necessary to increase the absorbed energy to the maximum level by developing the design of the UV light device with an appropriate lamp and size in order to achieve the desired effect.

5.5.3 Reactors

Reactors are devices used for UV light application to liquids. A UV reactor contains UV lamps inside. Each UV lamp is in a separate protective quartz tube to prevent direct contact with liquid. The liquid flowing through the UV reactor is exposed to UV rays emitted from lamps. Thus, the microorganisms in the liquid become ineffective. In the selection of UV reactors, the physical, chemical and microbiological properties of the liquid to be disinfected and the amount of the liquid passing through are the most important parameters. In this context, the UV light dose should be determined according to the nature of the fluid and the target microorganism (Koca et al., 2018). In addition, to increase the efficiency of disinfection, parameters such as sediment and turbidity in liquid should be removed with sensitive filters. The flow pattern of liquid in the UV reactor also has a significant effect on the total UV dose due to the differences in the position and residence times of the microorganisms in certain regions of the irradiated field. The inactivation of microorganisms increases using a turbulent flow in continuous flow UV reactors (Binot et al., 1998).

5.6 THE APPLICATIONS OF UV LIGHT IN THE DAIRY INDUSTRY

5.6.1 Disinfection of Air in the Production Area

Clean and fresh air is necessary for the food processing area. UV technology can be used for preventing the spread of airborne diseases by the inhibition of airborne pathogenic microorganisms in the field of production, packaging, cooling, storage and ripening. For this purpose, low-pressure mercury vapor lamps are successfully used as UV light sources. The efficiency of this process depends on the volume of the area and the power of the UV lamp (Keklik et al., 2012; Masotti et al., 2019; Donaghy et al. 2009).

5.6.2 Disinfection of Water Used During Processing

UV light has been used to disinfect water for several years and has become a successful process that eliminates several types of microorganisms (Crook et al 2014). UV-C technology is a good alternative to chlorine disinfection. In the dairy industry, it is possible to use UV systems for the disinfection of drinking water, process water, wastewater and brine.

5.6.3 Surface Applications of Packaging Materials and Equipment

5.6.3.1 Packaging Materials

In the food industry, the use of UV light for the decontamination of packaging material is becoming widespread. The number of microorganisms on the surfaces of packaging materials such as boxes, cartons, foils, films, wrappings, containers, bottles, caps, closures and lids can be reduced or eliminated by applying the appropriate UV light doses. The packages can be treated with UV light before filling or closing the lid, or the packaged food can be exposed to UV-C light. The effectiveness of UV treatment is better on smooth surfaces (Ansari and Datta, 2003).

5.6.3.2 Food Contact Surfaces

The cross-contamination of microorganisms from equipment to the products is an important issue in dairy technology. UV light can be used to provide disinfection of the surfaces of the conveyor and other equipment used in preparation and production as well as storage areas. For effective inhibition, microorganisms must be exposed to UV light directly. There should be no obstruction between the UV source and the surface to be sterilized. The success of this application also depends on the cleanliness of the material surfaces because dirt would absorb the radiation and hence protect the bacteria (Carrasco et al., 2012).

5.7 EFFICACY OF UV ON MARKET MILK

Some oxidative changes in raw milk for lipids and proteins are induced by unsaturated fatty acids, metal ions, oxidases and other pro-oxidants that are present in the milk (Hu et al., 2015). Dairy products are light-sensitive which decreases the nutritional benefit, including the content of unsaturated fatty acids and essential vitamins, particularly riboflavin and alpha-tocopherol (Westermann et al., 2009: Mortenasen et al., 2004). The thermal processes, that is, pasteurization and sterilization, inactivate microorganisms in dairy products. But as a result of these processes, nutrient and aroma loss, non-enzymatic browning and organoleptic differentiation especially in dairy products are observed. So, some alternative approaches are required to include microbial inactivation as non-thermal processes are triggered by high temperatures. Some methods such as ultraviolet radiation (UV), high pressure (HP), pulsed light (PL), supercritical carbon dioxide (SC-CO2) or pulsed electric field (PEF) are used in food. The reproduction of bacteria, viruses, fungi and other microorganisms is prevented when their DNA is exposed to UV radiation. At low or medium pressure, a mercury lamp emits a continuous beam in a mono- or polychromatic mode. UV is a non-thermal process that inactivates microorganisms in edible liquids such as apples, cranberry and grapefruit juices, milk, sugar solutions, lemons, water and other liquid foods and liquid eggs. According to the National Advisory Committee on microbiological criteria for foods, ultraviolet (UV) radiation can be safely used for the pasteurization of certain foods. In fruit juices and water, the bacteria and viruses can be inactivated at 200–280 nm UV wavelengths (Sastry et al., 2000; Hanes et al., 2002; Quintero-Ramos et al., 2004; Wright et al., 2000). The use of UV radiation for the treatment of water and food under specific conditions for market approval was given by the FDA (CFR, 2005). Smith et al. (2002) reported that "in principle, the bacterial content of milk can be adequately controlled by exposure to pulsed ultraviolet light." In practice, the application of UV treatment to milk has been challenging because the penetration of UV light into the liquid is limited by the solids content of milk which reduces its efficacy, and also, oxidation and sensory defects in milk can be caused by excessive exposure to UV (Smith et al., 2002). Very few researchers have concentrated on the effects of UV on a biochemical and chemical perspective of dairy products. Some authors concluded that the chemical composition of milk is not significantly affected by UV light application (Hu et al., 2015; Cappozzo et al., 2002). Likewise, Cilliers et al. (2014) concluded that most macro- and micro-components were not affected by the application of UV light to bovine milk, but the amount of cholesterol relative to pasteurized milk decreased. In goat milk samples, the TBARs and acid degree values increased in response to oxidative shifts in milk with UV light (Matak et al., 2007). With light exposure, the nutritional value and sensory qualities of dairy products can be altered based on lipids, protein oxidation and light sensitivity. UV-C therapy has the capacity to accelerate the formation of volatile compounds. In fact, by the application of UV light (at 254 nm, 11.8 W/m2), Hu et al. (2017) found an increase in the variety and content of volatile compounds of cow milk. The vitamin D content of pasteurized milk treated with UV radiation increased as suggested by the European Food Safety Authority (EFSA). In cow and goat milk the effects of UV light on vitamins A, B2, C and E have been tested by Guneser and Karagul (2012) and the susceptibility of vitamins to UV light in milk samples was determined to be C > E > A > B2. It was inferred that decrement in the content of vitamins and their reduction levels depend on the initial sum of vitamins and the number of machine passes. In comparison to other studies, Cappozzo et al. (2015) observed a decrease in vitamin D to undetectable levels during UV light, HTST and UHT processing of raw milk. Protein oxidation in dairy systems has a vital impact on protein properties and functionalities, while UV light can cause the degradation or modification of proteins which leads to changes in solubility, sensitivity to heat, mechanical properties and digestion by proteases. Reinemann et al. (2006) announced that raw cow milk under UV treatment produced more than a 3 log reduction in total bacterial numbers. The highest reduction was found for coliform bacteria followed by psychrotrophs, thermodurics and spore formers. Microbial counts of UV-treated milk were lower compared to those of control milk (Rossitto et al., 2012). UV-C treatment of raw cow milk was capable of reducing total viable count by 2.3 log (Bandla et al., 2012). UV light treatment in milk can be used as a method to reduce the number of psychrotrophic bacteria to prolong the storage period of cooled raw milk (Wright et al., 2000; Koutchma, 2009; Cilliers et al., 2014; Krishnamurthy et al., 2007; Krishnamurthy et al 2004).

REFERENCES

Ansari, I.A. and Datta, A.K. 2003. An overview of sterilization methods for packaging materials used in aseptic packaging systems. *Food and Bioproducts Processing*, 81(1): 57–65.

Bandla, S., Choudhary, R., Watson, D.G. and Haddock, J. 2012. UV-C treatment of soymilk in coiled tube UV reactors for inactivation of Escherichia coli W1485 and Bacillus cereus endospores. *LWT—Food Science and Technology*, 46(1): 71–76.

Binot, P., Omnium de traitements et de valorisation, O.T.V. 1998. *Reactor for UV Radiation for the Treatment of Liquids*. U.S. Patent 5,725,757.

Bintsis, T., Litopoulou-Tzanetaki, E. and Robinson, R.K. 2000. Existing and potential applications of ultraviolet light in the food industry—A critical review. *Journal of the Science of Food and Agriculture*, 80(6): 637–645.

Bolton, J.R. and Linden, K.G. 2003. Standardization of methods for fluence (UV dose) determination in bench-scale UV experiments. *Journal of Environmental Engineering*, 129(3): 209–215.

Butz, P. and Tauscher, B. 2002. Emerging technologies: Chemical aspects. *Food Research International*, 35(2–3): 279–284.

Cappozzo, J.C., Koutchma, T. and Barnes, G. 2015. Chemical characterization of milk after treatment with thermal (HTST and UHT) and nonthermal (turbulent flow ultraviolet) processing technologies. *Journal of Dairy Science*, 98(8): 5068–5079.

Carrasco, E., Morales-Rueda, A. and García-Gimeno, R.M. 2012. Cross-contamination and recontamination by Salmonella in foods: A review. *Food Research International*, 45(2): 545–556.

Choudhary, R. and Bandla, S. 2012. Ultraviolet pasteurization for food industry. *International Journal of Food Science and Nutrition Engineering*, 2(1): 12–15.

Cilliers, F.P., Gouws, P.A., Koutchma, T., Engelbrecht, Y., Adriaanse, C. and Swart, P. 2014. A microbiological, biochemical and sensory characterisation of bovine milk treated by heat and ultraviolet (UV) light for manufacturing Cheddar cheese. *Innovative Food Science and Emerging Technologies*, 23: 94–106.

Crook, J.A., Rossito, P.V., Parko, J., Koutchma, T. and Cullor, J.S. 2014. Efficacy of ultraviolet (UV-C) light in a thin-film turbulent flow for the reduction of milk borne pathogens. *Foodborne Pathogens and Disease*, 12: 506–513.

Delorme, M.M., Guimarães, J.T., Coutinho, N.M., Balthazar, C.F., Rocha, R.S., Silva, R., Margalho, L.P., Pimentel, T.C., Silva, M.C., Freitas, M.Q., Granato, D., Sant'Ana, A.S., Duart, M.C.K.H. and Cruz, A.G. 2020. Ultraviolet radiation: An interesting technology to preserve quality and safety of milk and dairy foods. *Trends in Food Science and Technology*, 102: 146–154.

Donaghy, J., Keyser, M., Johnston, J., Cilliers, F.P., Gouws, P.A. and Rowe, M.T. 2009. Inactivation of Mycobacterium avium ssp. paratuberculosis in milk by UV treatment. *Letters in Applied Microbiology*, 49(2): 217–221.

Dunn, J., Ott, T. and Clark, W. 1995. Pulsed light treatment of food and packaging. *Food Technology*, 49: 95–98.

FDA, U. S. Food and Drug Administration. Ultraviolet (UV) radiation. Available in: https://www.fda.gov/radiation-emitting-products/tanning/ultraviolet-uvradiation#1. Accessed date: 4 October 2019.

Franz, C.M.A.P., Specht, I., Cho, G.S., Graef, V. and Stahl, M.R. 2009. UV-C-inactivation of microorganisms in naturally cloudy apple juice using novel inactivation equipment based on Dean vortex technology. *Food Control*, 20(12): 1103–1107.

Gomez Lopez, V.M., Koutchma, T. and Linden, K. 2012. Ultraviolet and pulsed light processing of fluid foods. In: Cullen, P. J., Tiwari, B., Valdramidis, V. editors. *Novel Thermal and Nonthermal Technologies for Fluid Foods*. San Diego: Academic Press, pp. 185–223.

Gómez-López, V.M., Ragaert, P., Debevere, J. and Devlieghere, F. 2007. Pulsed light for food decontamination: A review. *Trends in Food Science and Technology*, 18(9): 464–473.

Guerrero-Beltrán, J.A. and Barbosa-Cánovas, G.V. 2004. Advantages and limitations on processing foods by UV Light. *Food Science and Technology International*, 10(3): 137–147.

Guneser, O. and Karagul Yuceer, Y.Y. 2012. Effect of ultraviolet light on water- and fat-soluble vitamins in cow and goat milk. *Journal of Dairy Science*, 95(11): 6230–6241.

Ha, J.W., Back, K.H., Kim, Y.H. and Kang, D.H. 2016. Efficacy of UV-C irradiation for inactivation of foodborne pathogens on sliced cheese packaged with different types and thicknesses of plastic films. *Food Microbiology*, 57: 172–177.

Hanes, D.E., Worobo, R.W., Orlandi, P.A., Burr, D.H., Miliotis, M.D., Robl, M.G., Bier, J.W., Arrowood, M.J., Churey, J.J. and Jackson, G.J. 2002. Inactivation of Cryptospridium parvum oocysts in fresh apple cider by UV irradiation. *Applied and Environmental Microbiology*, 68(8): 4168–4172.

Haug, A., Høstmark, A.T. and Harstad, O.M. 2007. Bovine milk in human nutrition—A review. *Lipids in Health and Disease*, 6(1): 1–16.

Hu, G., Zheng, Y., Liu, Z. and Deng, Y. 2017. Effects of UV-C and single- and multiple-cycle high hydrostatic pressure treatments on flavor evolution of cow milk: Gas chromatography mass spectrometry, electronic nose, and electronic tongue analyses. *International Journal of Food Properties*, 20(7): 1677–1688.

Hu, G., Zheng, Y., Wang, D., Zha, B., Liu, Z. and Deng, Y. 2015. Comparison of microbiological loads and physicochemical properties of raw milk treated with single-/multiple-cycle high hydrostatic pressure and ultraviolet-C light. *High Pressure Research*, 35(3): 330–338.

Kasahara, I., Carrasco, V. and Aguilar, L. 2015. Inactivation of *Escherichia coli* in goat milk using pulsed ultraviolet light. *Journal of Food Engineering*, 152: 43–49.

Keklik, N.M., Krishnamurthy, K. and Demirci, A. 2012. Microbial decontamination of food by ultraviolet (UV) and pulsed UV light. In: *Microbial decontamination in the food industry*. Woodhead Publishing. pp. 344–369.

Koca, N., Urgu, M. and Saatli, T. 2018. Ultraviolet light applications in dairy processing. In: KOCA (Ed.). *Technological Approaches for Novel Applications in Dairy Processing*. Intech Open (Chapter 1). DOI: 10.5772/intechopen.74291.

Koutchma, T., Keller, S., Chirtel, S. and Parisi, B. 2004. Ultraviolet disinfection of juice products in laminar and turbulent flow reactors. *Innovative Food Science and Emerging Technologies*, 5(2): 179–189.

Koutchma, T. 2009. Advances in ultraviolet light technology for non-thermal processing of liquid foods. *Food and Bioprocess Technology*, 2(2): 138–155.

Koutchma, T. 2019. *Ultraviolet Light in Food Technology: Principles and Applications* (Vol. 2). CRC Press.

Koutchma, T., Forney, L.J. and Moraru, C.I. 2009. *Ultraviolet Light in Food Technology: Principles and Applications*. Boca Raton, FL: Taylor & Francis Group, CRC Press. p. 278.

Koutchma, T., Orlowska, M. and Zhu, Y. 2012. *UV Light for Fruits and Fruit Products*. 93 Stone Rd West, Guelph: Agriculture and Agri-Food Canada Guelph, Food Research Center.

Koutchma, T., Popović, V. and Green, A. 2019. Chapter 1. Overview of Ultraviolet (UV) LEDs Technology for Applications in Food Production. In: *Ultraviolet LED Technology for Food Applications*. Academic Press. pp. 1–23.

Kowalski, W. 2009. *UVGI Disinfection Theory. Ultraviolet Germicidal Irradiation Handbook. UVGI for Air and Surface Disinfection*. New York: Springer. pp. 17–50.

Krishnamurthy, K., Demirci, A. and Irudayaraj, J.M. 2004. Inactivation of *Staphylococcus aureus* by pulsed UV-light sterilization. *Journal of Food Protection*, 67(5): 1027–1030.

Krishnamurthy, K., Demirci, A. and Irudayaraj, J.M. 2007. Inactivation of *Staphylococcus aureus* in milk using flow-through pulsed UV-light treatment system. *Journal of Food Science*, 72(7): M233–M239.

López, M.A., Palou, E., Barbosa, C.G., Tapia, M.S. and Cano, M.P. 2005. Ultraviolet light and food preservation. *Novel Food Processing Technologies*: 405–421.

Masotti, F., Cattaneo, S., Stuknytė, M. and De Noni, I. 2019. Airborne contamination in the food industry: An update on monitoring and disinfection techniques of air. *Trends in Food Science and Technology*, 90: 147–156.

Matak, K.E., Sumner, S.S., Duncan, S.E., Hovingh, E., Worobo, R.W., Hackney, C.R. and Pierson, M.D. 2007. Effects of ultraviolet irradiation on chemical and sensory properties of goat milk. *Journal of Dairy Science*, 90(7): 3178–3186.

Miller, R.V., Jeffrey, W., Mitchell, D. and Elasri, M. 1999. Bacterial responses to ultraviolet light. *ASM News-American Society for Microbiology*, 65: 535–541.

Mortensen, G., Bertelsen, G., Mortensen, B.K. and Stapelfeldt, H. 2004. Light-induced changes in packaged cheeses—A review. *International Dairy Journal*, 14(2): 85–102.

NASA, National Aeronautics and Space Administration 2019. The electromagnetic spectrum. Available in: https://imagine.gsfc.nasa.gov/science/toolbox/emspectrum1.html. Accessed Date: 4 October 2019.

Ortega-Rivas, E. 2012. Pulsed light technology. In: *Non-Thermal Food Engineering Operations*. Boston, MA: Springer. pp. 263–273. DOI: https://doi.org/10.1007/978-1-4614-2038-5.

Quintero-Ramos, A., Churey, J.J., Hartman, P., Barnard, J. and Worobo, R.W. 2004. Modeling of Escherichia coli inactivation by UV irradiation at different pH values in apple cider. *Journal of Food Protection*, 67(6): 1153–1156.

Reinemann, D.J., Gouws, P., Cilliers, T., Houck, K. and Bishop, J. 2006. New methods for UV treatment of milk for improved food safety and product quality. American Society of Agricultural and Biological Engineers, ASABE Annual Meeting, 21493: Paper number 066088.

Rossitto, P.V., Cullor, J.S., Crook, J., Parko, J., Sechi, P. and Cenci-Goga, B.T. 2012. Effects of UV irradiation in a continuous turbulent flow UV reactor on microbiological and sensory characteristics of cow's milk. *Journal of Food Protection*, 75(12): 2197–2207.

Sastry, S.K., Datta, A.K. and Worobo, R.W. 2008. Ultra-violet light. *Journal of Food Science Supplement*, 65: 90–92.

Schalk, S., Adam, V., Arnold, E., Brieden, K., Voronov, A. and Witzke, H.D. 2006. UV-lamps for disinfection and advanced oxidation—Lamp types, technologies and applications. *IUVA News*, 8(1): 32–37.

Schubert, E.F. 2006. *Light-Emitting Diodes*, 2nd edition. New York, NY: Cambridge University Press. DOI: https://doi.org/10.1017/CBO9780511790546.

Shin, J.Y., Kim, S.J., Kim, D.K. and Kang, D.H. 2016. Fundamental characteristics of deep-UV light-emitting diodes and their application to control food borne pathogens. *Applied and Environmental Microbiology*, 82(1): 2–10.

Smith, W.L., Lagunas-Solar, M.C. and Cullor, J.S. 2002. Use of pulsed ultraviolet laser light for the cold pasteurization of bovine milk. *Journal of Food Protection*, 65(9): 1480–1482.

Westermann, S., Brüggemann, D.A., Olsen, K. and Skibsted, L.H. 2009. Light-induced formation of free radicals in cream cheese. *Food Chemistry*, 116(4): 974–981.

Woldemariam, H.W. and Emire, S.A. 2019. High pressure processing of foods for microbial and mycotoxins control: Current trends and future prospects. *Cogent Food and Agriculture*, 5(1): 1622184.

Wright, J.R., Sumner, S.S., Hackney, C.R., Pierson, M.D. and Zoecklein, B.W. 2000. Efficacy of ultraviolet light for reducing Escherichia coli O157:H7 in unpasteurized apple cider. *Journal of Food Protection*, 63(5): 563–567.

Chapter 6

Application of Ultrasound for Dairy Product Processing

Maryam Enteshari, Collette Nyuydze and Sergio I. Martinez-Monteagudo

CONTENTS

6.1	Introduction	81
6.2	Principles of Ultrasound	82
6.3	Applications of Ultrasound in Dairy Manufacture	84
	6.3.1 Emulsification	84
	6.3.2 Homogenization	85
	6.3.3 Rise in the Product Temperature	85
	6.3.4 Reduction of Microbial Load	87
	6.3.5 Sonocrystalization	87
	6.3.6 Solubility of Dairy Powders	87
	6.3.7 Foaming	88
	6.3.8 Heat Stability	88
	6.3.9 Gelation of Milk Proteins	88
	6.3.10 Fermentation	89
6.4	Conclusions	89
References		89

6.1 INTRODUCTION

Over the past few decades, a number of dairy manufacturing operations have been accomplished through the use of ultrasound. Reduction of particle size, emulsification, encapsulation, denaturation of proteins, crystallization of lactose, inactivation of enzymes, and reduction of the microbial load are examples of such operations.

Ultrasound-induced modifications of milk constituents have led to a growth in the number of patents and scientific publications in the field of dairy manufacture. This chapter provides an overview of the principles of ultrasound, a summary of the ultrasound-induced properties of milk constituents, and an overview of relevant dairy manufacturing applications, including emulsification, reduction of microbial load, crystallization, foaming, and gelation.

6.2 PRINCIPLES OF ULTRASOUND

In a nutshell, ultrasound refers to the generation and use of inaudible acoustic waves within the frequency range of 20 kHz to 10 MHz (Gallego-Juarez & Graff, 2015). An ultrasonic system consists of a source of electrical power and a transducer that converts the electrical energy into ultrasonic waves. A number of transducers have been used in ultrasonic systems, including piezoelectric and magnetostrictive. Fundamentals of the transducers and their behavior in frequency, time, and temperature are discussed elsewhere (Pardo, 2015). Figure 6.1 illustrates the sequence of ultrasonic events in a typical ultrasound treatment. Fundamentals and developments of the generation and transmission of ultrasonic waves can be found in classical textbooks (Richardson, 1962; Cheeke, 2012; Gallego-Juarez & Graff, 2015). Compressed waves generated by the transducer are transmitted by a medium (solid, liquid, or gas) into the end material (for instance, food). In dairy manufacturing, the medium and end material are dairy products (cheese, yogurt, milk, etc.) or milk constituents (lactose, fat, proteins, etc.). During the ultrasonic event, compressed waves interact with both the medium and end material, creating several fluid dynamics phenomena, including microstreaming, microjet, and shockwaves. The extent to which these processes occur controls the final application of ultrasound.

The ultrasound applications are further divided according to the frequency used: low-frequency/high-intensity and high-frequency/low-intensity applications. The distinction between low and high intensity is somewhat ambiguous since the limit between them depends on the characteristics of the medium (Khadhraoui et al., 2019). Figure 6.2 illustrates the frequency-intensity domain of ultrasound and some common applications.

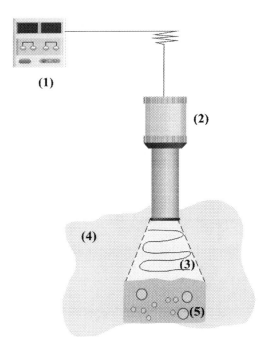

Figure 6.1 Illustration of the sequence of events in a typical ultrasound application. (1) Power source; (2) transducer; (3) propagation of waves; (4) medium; (5) end material.

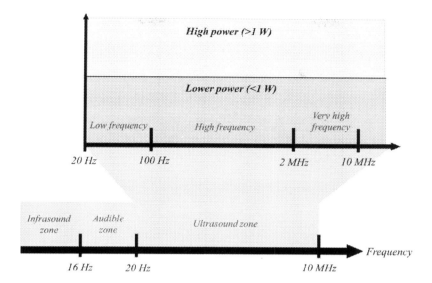

Figure 6.2 Illustration of ultrasound frequency and power range. Source: reprinted from Khadhraoui, B., Fabiano-Tixier, A.-S., Robinet, P., Imbert, R., & Chemat, F. (2019). "Ultrasound technology for food processing, preservation, and extraction." In *Green Food Processing Techniques*, with permission from Elsevier.

Applications within the high-frequency/low-intensity spectrum consist of nondestructive testing and medical imaging. In these applications, the power applied to the transducer is low (milliwatt), while the resulting frequencies are very high (megahertz). Such power levels are not enough to induce modification in the medium and end material. Moreover, high-frequency/low-intensity waves propagate through the medium with a constant speed and linear acoustics (Sapozhnikov, 2015).

On the other hand, low-frequency/high-intensity applications are quite diverse since the power used promotes physical and chemical changes within the medium and end material. Welding of metals, particle dispersion, cleaning, emulsification, degassing, defoaming, drying, and extraction are examples of low-frequency/high-intensity applications (Gallego-Juarez & Graff, 2015). These applications are collectively referred to as power ultrasound, where power levels applied range from 10 to 1,000 s of watts and frequencies in the order of 10 to 500 kHz. In power ultrasound, low-frequency/high-intensity waves propagate through a medium in a non-linear fashion, where a series of mechanisms may be triggered, including wave distortion, streaming, cavitation in liquids, microjet, and shockwaves (Sapozhnikov, 2015).

In power ultrasound, the waves are characterized by four parameters, including frequency (f), energy density (ε), ultrasonic power (P), and acoustic intensity (I_a). Table 6.1 shows the main parameters used to characterize the ultrasonic waves.

A common effect of power ultrasound is the development of acoustic cavitation within the medium and end material. During the application of power ultrasound, a liquid experiences cycles of positive and negative pressure. Air bubbles are formed when the negative pressure of the liquid exceeds the ambient pressure, a process known as acoustic cavitation which is characterized by oscillating movements of energy. Overall, cavitation refers to the phenomenon of the formation of cavities or bubbles within a liquid, their growth, and subsequent collapse when the pressure of the liquid is suddenly reduced below its vapor pressure (Sim et al., 2021). Cavitation is

TABLE 6.1 PARAMETERS COMMONLY USED TO CHARACTERIZE THE ULTRASONIC WAVES

Parameter	Symbol and Units	Equation	Description
Frequency	$f\ [=]\ Hz$	N.E.A.	Number of cycles
Energy density	$\varepsilon\ [=]\ kJ\ cm^{-3}$	$\varepsilon = \dfrac{Power \cdot time}{Volume}$	Amount of energy provided per unit of volume
Ultrasonic power	$P\ [=]\ J\ s^{-1}$	$P = m \cdot C_p \cdot \left(\dfrac{dT}{dt}\right)$	Amount of energy dissipated as heat
Acoustic intensity	$I_a\ [=]\ W\ cm^2$	$I_a = \dfrac{P}{S_a}$	Acoustic energy emitted from the transducer

N.E.A. – no equation associated; m – mass of sample; C_p – heat capacity of the sample; $\dfrac{dT}{dt}$ – temperature rise per second; S_a – surface area of the transducer.

somewhat similar to the evaporation process, where latent heat is supplied from the surroundings. However, the oscillatory nature of acoustic cavitation creates important physical and chemical processes. Collapsing mechanisms of cavitation bubbles have been reviewed elsewhere (van Wijngaarden, 2016). In general, the collapse of the cavitation bubbles occurs within a very short time span (milli- or microseconds), resulting in shockwaves that raise the liquid temperature.

6.3 APPLICATIONS OF ULTRASOUND IN DAIRY MANUFACTURE

6.3.1 Emulsification

Emulsification is a key processing step in manufacturing a number of dairy and food products, including ice cream, infant formulas, sauces, dressings, soups, mayonnaise, butter, and margarine (Martinez-Monteagudo et al., 2017). Emulsification involves the dispersion of two immiscible liquids within a continuous medium, stabilized by the addition of a surfactant and application of mechanical energy (Nyuydze & Martínez-Monteagudo, 2021). The dispersion of immiscible liquids is thermodynamically unstable by nature, and separation of the liquids may occur over time. The preparation of stable emulsions requires a high input of mechanical energy to break the liquid interfaces. In ultrasound processing, sound waves are transmitted through the liquid at frequencies above the human hearing threshold (> 16 kHz), resulting in compression and stretching of the molecular spacing leading to cavitation bubbles. Upon collapsing, these bubbles release energy in the form of heat, shockwaves, and shearing, which can be put to work for dispersion, mixing, and emulsification (Chandrapala & Leong, 2015). Emulsification through ultrasound is characterized by the collapse of the bubbles at or near the oil–water interface that disrupts and mixes the two phases, forming fine droplets (Modarres-Gheisari et al., 2019). Ultrasound has been shown to produce stable oil-in-water emulsions over a wide range of oil content, 3–20%, v/v. Aslan and Dogan (2018) emulsified olive oil (7–15%, v/v) in reconstituted skim milk by the application of ultrasound treatment (24 kHz for 3 min). The resulting emulsions were stable against creaming without the addition of surfactants. Similarly, Kaci et al. (2014) dispersed vegetable oil (5–15%, v/v) in water without the addition of surfactants using high-frequency ultrasound generated by a piezoelectric ceramic transducer. The application of ultrasound treatment (100 W for 8 min) emulsified black seed oil (7%, v/v) in skim milk, and the emulsions were stable for eight days at 4° C without the addition of surfactants. An investigation of the emulsification of flaxseed oil (7–21%, v/v) showed that the ultrasound treatment (20 kHz for up to 8 min) dispersed droplets of oil in skim milk, and such droplets were stable for nine days at 4° C (Anandan et al., 2017). In summary, the literature on ultrasound emulsification appears to produce stable oil-in-water emulsions with reduced surfactants.

6.3.2 Homogenization

Homogenization is the physical process of subdividing large particles into a large number of smaller particles having a small size (Martinez-Monteagudo et al., 2017). The subdivision process involves the disruption of the large particles and the dispersion of the newly created particles. Ultrasound represents a technological alternative for milk homogenization since it reduces the average size of milk fat due to cavitation (Chandrapala et al., 2016). Cavitation refers to the formation, growth, and collapse of cavities within a liquid. Overall, cavitation causes significant thermo-mechanical effects, resulting in shockwaves that raise the liquid temperature and provide the energy for the particle break-up. Homogenization using ultrasound breaks up the milk fat globule membrane, creating smaller globules. Villamiel and de Jong (2000) reported a unimodal distribution of fat globules with an average size of 0.57–0.95 µm after sonication of raw milk. Ultrasound treatment of 20 kHz and 55° C for 10 min yielded values of the average size of fat globules comparable to those obtained in conventional homogenization (< 1.0 µm).

6.3.3 Rise in the Product Temperature

The application of ultrasound in different liquids results in the transmission of sound waves (as longitudinal waves), enabling the formation of rapidly growing bubbles that expand during negative pressure excursion and collapse during positive excursion, increasing the temperature, pressure, and shear forces of the medium (Jambrak et al., 2009). The magnitude of the ultrasound (mechanical energy) is subsequently dissipated partly as heat when ultrasound passes through the material; hence the temperature is recorded as a function of time, resulting in an estimation of power in watts (Jambrak et al., 2009).

Figure 6.3 shows the graph of temperature change with time at ultrasound treatments of 50 to 100% amplitude for water. The temperatures increase with increasing time, and amplitude is linear. This trend shows that increasing amplitude results in an increase in power transmitted through the sample. This results in the rapid formation and breakdown of bubbles that in turn increase the temperature of the samples.

Figure 6.4 shows the temperature change with different material compositions. This relates to threshold values reached before any physical or chemical changes occur due to ultrasound intensities on different material compositions. The impact of the ultrasound on material composition shows significant differences for samples with 39.5% total solids and 43.5% fat content, treated at 100% amplitude. The effect of sonication can vary depending on the experimental conditions such as acoustic power density, the volume of the sample, the temperature of the solution, and other factors (Shanmugam et al., 2012). The magnitude at which ultrasound travels through the material is dependent on the intensity applied and the type of material treated. Ultrasound effects on the kinetics of mass transport appear only when an acoustic intensity threshold is reached. The

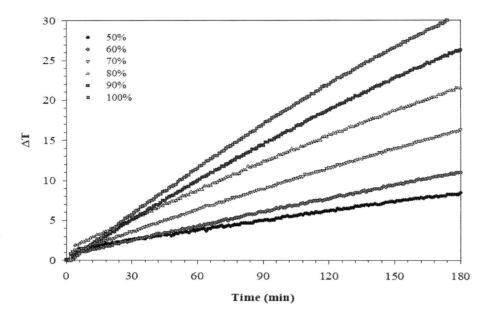

Figure 6.3 Temperature modeling at amplitudes 50, 60, 70, 80, 90, and 100%.

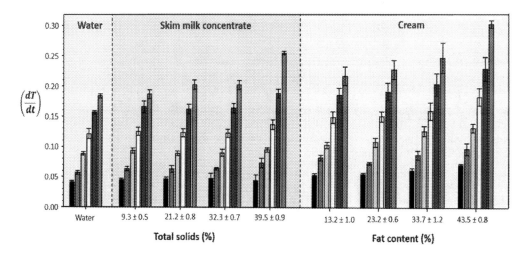

Figure 6.4 Effect of temperature change with material compositions.

effects of ultrasound on the model systems prepared by ultrasonication of water and different material compositions became apparent when threshold values were reached before chemical or biochemical changes occurred. Measuring the variation of ultrasound properties enables the generation of information about the properties of the system. Overall, different ultrasound treatments and processing times will result in functional changes in different materials depending on their compositions and threshold limits.

6.3.4 Reduction of Microbial Load

The increase in the sample temperature due to sonication has been used as a technological tool to enhance the thermal treatments and reduce the microbial load. Interestingly, the combination of ultrasound and heat is known as thermosonication, enhancing lethality compared with traditional pasteurization (Chandrapala & Zisu, 2018). The amplitude of ultrasonic waves, treatment time, the volume of food being processed, food composition, and microorganism type remarkably influence the final reduction level. Overall, gram-positive bacteria are more susceptible to sonication than gram-negative. The inactivation of microorganisms through ultrasound is thought to induce irreversible damage in the cell walls, affecting the structural and functional components of the cells, leading to cell lysis (Bermudez-Aguirre et al., 2011). A comprehensive review of the effect of ultrasound on the inactivation of the microorganism of relevance for the dairy industry can be found elsewhere (Silva & Chandrapala, 2021).

6.3.5 Sonocrystalization

Lactose is industrially produced from the whey produced by the cheese and casein industry. The production of lactose from solution involves concentration, nucleation, and crystal growth, washing, centrifugation, drying, and packaging. The manufacturing steps of lactose are general knowledge, and specific details of the individual operations can be found elsewhere (Paterson, 2017). Lactose is commercially available as edible-grade, anhydrous, or pharmaceutical-grade, depending on subsequent purification steps. Lactose is primarily used in pharmaceuticals, infant formula, and confectionary markets (Cheng & Martinez-Monteagudo, 2019). Overall, supersaturation, nucleation, and crystal growth are the key processing steps during lactose manufacturing. Crystallization protocols consist of a prescribed time, temperature, and cooling rate known to give a desirable yield of lactose crystals. Silva and Chandrapala (2021) summarized the literature dealing with the sonocrystallization of lactose. Ultrasound can enhance the overall crystallization process by accelerating the nucleation due to the action of acoustic waves (Bund & Pandit, 2007). Additionally, sonocrystallization reduces the crystal size and narrow distribution, decreased agglomeration, and increased uniformity (Patel & Murphy, 2009). Mechanistically, shockwaves generated through sonication increase the number of collisions, increasing the number of nuclei during the nucleation step and eventually a greater number of nuclei. An investigation of the sonocrystallization of lactose showed that a power intensity of 0.46 W g^{-1} accelerated more than ten times the nucleation rate of traditional stirring (Dincer et al., 2014). Zisu et al. (2014) sonocrystallized lactose from concentrated whey using an ultrasound treatment of 20 kHz and reported a narrower size distribution compared with traditional crystallization.

6.3.6 Solubility of Dairy Powders

Ultrasound has been used to improve the solubility of dairy powders since it not only produces smaller particles by disrupting protein aggregates but also exposes the hydrophilic amino acids within the protein. Ultrasound produced highly localized shear forces that break up larger particles, increasing the surface area of the powder particles. Milk protein concentrate was ultrasonicated at an energy density of 153 J mL^{-1}, and an improvement in solubility was observed compared with the untreated powder (Chandapala et al., 2014). Zhang et al. (2018) sonicated micellar casein concentrate for 5 min at an energy density of 58 W L^{-1} and found an increase in the solubility up to 90%. Sonication (20 and 40 kHz) of whey protein isolate improved the solubility at 10% reconstitution due to the changes in the three-dimensional structure of the proteins (Jambrak et al., 2008). Chandrapala et al. (2014) used ultrasound at an energy density of 31 W L^{-1} prior to spray drying to improve the solubility of milk protein concentrate. These authors reported that an ultrasound treatment disrupted the aggregation of proteins without changing the surface properties of the powder.

6.3.7 Foaming

Ultrasound has also improved the foaming ability of milk protein due to a partial unfolding induced by the localized shear forces. Jambrak et al. (2008) increased the foaming ability of whey protein isolate using high-intensity ultrasound (40 kHz). These authors reported that ultrasound evenly dispersed protein and fat particles. A sonication treatment consisting of 400 W for up to 25 min was applied to 20–60% whey protein suspensions to improve the foaming ability (Tan et al., 2016). These authors suggested that a sonicated-whey protein can substitute for egg proteins during baking. Similarly, the foaming properties and the viscosity and consistency index of whey protein concentrate were improved with an ultrasound treatment (Tan et al., 2015).

6.3.8 Heat Stability

Concentration of milk is a common operation aimed at increasing nutrient density and lowering transportation costs. Concentrated milk possesses low heat stability, and it is not suitable for sterilization treatments. Ashokkumar et al. (2009) pre-heated a whey protein solution and then sonicated it to improve the heat stability of the whey protein while controlling the excessive viscosity during heat treatment. Chandrapala et al. (2013) postulated that ultrasound disrupted large protein aggregates and lowered the surface hydrophobicities, preventing aggregation during heating. The addition of calcium is a common practice in order to improve the heat stability of milk concentrates. An investigation of the effect of ultrasound and the addition of calcium in skim milk showed a similar viscosity to the untreated control, while the heat stability was significantly improved.

6.3.9 Gelation of Milk Proteins

Ultrasound has also been used prior to acid, rennet, or heat gelation of milk protein (Silva & Chandrapala, 2021). Ultrasound prior to acid gelation has been illustrated by Nguyen and Anema (2017), who sonicated (22 kHz frequency and 50 W output power) whole milk followed by acidification. These authors reported that the obtained gels displayed higher firmness compared to untreated milk. Sonication exposes the surface of unfolded whey proteins associated with fat globules, causing irreversible denaturation of whey proteins in whole milk. Hennelly et al. (2006) produced set-type yogurt from sonicated milk (24 kHz and 400 W), and the resulting yogurt exhibited increased firmness and viscosity compared with the untreated yogurts. Another application of ultrasound is prior to heat-induced gels, where a three-dimensional structure is created upon aggregation of denatured proteins. A sonication treatment of 13–50 W increased the gel strength and reduced the syneresis of heat-induced gels compared with the control samples (Zisu et al., 2011). Ultrasound has also been used for rennet gelation during the manufacture of cheese (Silva & Chandrapala, 2021). An investigation on the effect of ultrasound on the renneting properties of milk showed that ultrasound treatment (20 kHz) improved the curd firmness and gel network compared with untreated samples.

6.3.10 Fermentation

Fermentation extends the shelf life of milk products while improving digestibility and enhancing nutritional value. An investigation on the application of ultrasound at an energy density of 17.2 kW/m^2 prior to fermentation showed a less viable cell count than a non-sonicated sample (Wang & Sakakibara, 1997). Additionally, sonication triggered the diffusion of galactosidase from lactic acid bacteria cells expanding the galactosidase activity. Barukčić et al. (2015) studied the effect of thermosonication (480–600 W and 45–55° C) on whey improvement prior to the fermentation. Sonication increased membrane permeability from the cells that facilitate the release of intracellular enzyme β-galactosidase, improving the overall quality of the fermented product. Ultrasound treatment of milk (480 W and 55° C for 8 min) increased the viable cell counts with improved sensory characteristics (Chandrapala & Sizu, 2018). These results suggested that sonication can positively alter product quality and enhance the formation of microbial biomass.

6.4 CONCLUSIONS

Intelligent combinations of ultrasound frequency and intensity have proven to be an alternative technique for performing a number of operations for the manufacture of dairy products and ingredients, including emulsification, homogenization, inactivation of enzymes and pathogens, and modification functionality of proteins, extraction, cleaning, and others. Despite the research efforts summarized in this chapter, the localized nature of the technology has been a hurdle for broader adaptation for high-throughput products and ingredients. Nevertheless, the significant progress made in the field of non-linear acoustics over the past decade has created promising developments in the way ultrasound waves are dissipated and utilized.

REFERENCES

Anandan, S., Keerthiga, M., Vijaya, S., Asiri, A. M., Bogush, V., & Krasulyaa, O. (2017). Physicochemical characterization of black seed oil-milk emulsions through ultrasonication. *Ultrasonics Sonochemistry*, 38, 766–771.

Ashokkumar, M., Lee, J., Zisu, B., Bhaskarcharya, R., Palmer, M., & Kentish, S. (2009). Hot topic: Sonication increases the heat stability of whey proteins. *Journal of Dairy Science*, 92(11), 5353–5356.

Aslan, D., & Dogan, M. (2018). The influence of ultrasound on the stability of dairy-based, emulsifier-free emulsions: Rheological and morphological aspect. *European Food Research and Technology*, 244(3), 409–421.

Barukčić, I., Lisak Jakopović, K. L., Herceg, Z., Karlović, S., & Božanić, R. (2015). Influence of high intensity ultrasound on microbial reduction, physico-chemical characteristics and fermentation of sweet whey. *Innovative Food Science and Emerging Technologies*, 27, 94–101.

Bermúdez-Aguirre, D., Mobbs, T., & Barbosa-Cánovas, G. V. (2011). Ultrasound applications in food processing. In: Editor(s): Hao Feng, Gustavo V. Barbosa-Cánovas, Jochen Weiss, *Ultrasound Technologies for Food and Bioprocessing*, Springer-Verlag, New York, 65–105.

Bund, R. K., & Pandit, A. B. (2007). Sonocrystallization: Effect on lactose recovery and crystal habit. *Ultrasonics Sonochemistry*, 14(2), 143–152.

Chandrapala, J., & Leong, T. (2015). Ultrasonic processing for dairy applications: Recent advances. *Food Engineering Reviews*, 7(2), 143–158.

Chandrapala, J., Ong, L., Zisu, B., Gras, S. L., Ashokkumar, M., & Kentish, S. E. (2016). The effect of sonication and high pressure homogenisation on the properties of pure cream. *Innovative Food Science and Emerging Technologies*, 33, 298–307.

Chandrapala, J., & Zisu, B. (2018). Ultrasound technology in dairy processing, In: *Ultrasound Technology in Dairy Processing*. Springer Briefs in Molecular Science. Springer, Cham. https://doi.org/10.1007/978-3-319-93482-2_1.

Chandrapala, J., Zisu, B., Kentish, S., & Ashokkumar, M. (2013). Influence of ultrasound on chemically induced gelation of micellar casein systems. *The Journal of Dairy Research*, 80(2), 138.

Chandrapala, J., Zisu, B., Palmer, M., Kentish, S. E., & Ashokkumar, M. (2014). Sonication of milk protein solutions prior to spray drying and the subsequent effects on powders during storage. *Journal of Food Engineering*, 141, 122–127.

Cheeke, D. J. (2012). *Fundamentals and Applications of Ultrasonic Waves*. 2nd edition, CRC Press, Boca Raton, FL.

Cheng, S., & Martínez-Monteagudo, S. I. (2019). Hydrogenation of lactose for the production of lactitol. *Asia-Pacific Journal of Chemical Engineering*, 14(1), e2275.

Dincer, T. D., Zisu, B., Vallet, C. G. M. R., Jayasena, V., Palmer, M., & Weeks, M. (2014). Sonocrystallisation of lactose in an aqueous system. *International Dairy Journal*, 35(1), 43–48.

Gallego-Juarez, J., & Graff, K. (2015). *Power Ultrasonics Applications of High-Intensity Ultrasound*. 1st edition, Woodhead Publishing. ISBN 978-1-84569-989-5.

Hennelly, P. J., Dunne, P. G., O'sullivan, M., & O'riordan, E. D. (2006). Textural, rheological and microstructural properties of imitation cheese containing inulin. *Journal of Food Engineering*, 75(3), 388–395.

Jambrak, A. R., Lelas, V., Mason, T. J., Krešić, G., & Badanjak, M. (2009). Physical properties of ultrasound treated soy proteins. *Journal of Food Engineering*, 93(4), 386–393.

Jambrak, A. R., Mason, T. J., Lelas, V., Herceg, Z., & Herceg, I. L. (2008). Effect of ultrasound treatment on solubility and foaming properties of whey protein suspensions. *Journal of Food Engineering*, 86(2), 281–287.

Kaci, M., Meziani, S., Arab-Tehrany, E., Gillet, G., Desjardins-Lavisse, I., & Desobry, S. (2014). Emulsification by high frequency ultrasound using piezoelectric transducer: Formation and stability of emulsifier free emulsion. *Ultrasonics Sonochemistry*, 21(3), 1010–1017.

Khadhraoui, B., Fabiano-Tixier, A.-S., Robinet, P., Imbert, R., & Chemat, F. (2019). Chapter 2.Ultrasound technology for food processing, preservation, and extraction. In: Editor(s): Farid Chemat, Eugene Vorobiev, *Green Food Processing Techniques*, Academic Press, 23–56.

Martínez-Monteagudo, S. I., Yan, B., & Balasubramaniam, V. M. (2017). Engineering process characterization of high-pressure homogenization—From laboratory to industrial scale. *Food Engineering Reviews*, 9(3), 143–169.

Modarres-Gheisari, S. M. M., Gavagsaz-Ghoachani, R., Malaki, M., Safarpour, P., & Zandi, M. (2019). Ultrasonic nano-emulsification—A review. *Ultrasonics Sonochemistry*, 52, 88–105.

Nguyen, N. H., & Anema, S. G. (2017). Ultrasonication of reconstituted whole milk and its effect on acid gelation. *Food Chemistry*, 217, 593–601.

Nyuydze, C., & Martínez-Monteagudo, S. I. (2021). Role of soy lecithin on emulsion stability of dairy beverages treated by ultrasound. *International Journal of Dairy Technology*, 74(1), 84–94.

Pardo, L. (2015). Editor(s): Juan A. Gallego-Juárez, Karl F. Graff. *Piezoelectric Ceramic Materials for Power Ultrasonic Transducers, Power, Ultrasonics*, Woodhead Publishing, 101–125. ISBN 978-1-84569-989-5.

Patel, S. R., & Murthy, Z. V. P. (2009). Ultrasound assisted crystallization for the recovery of lactose in an anti-solvent acetone. *Crystal Research and Technology: Journal of Experimental and Industrial Crystallography*, 44(8), 889–896.

Paterson, A. H. J. (2017). Lactose processing: From fundamental understanding to industrial application. *International Dairy Journal*, 67, 80–90.

Richardson, E. G. (1962). *Ultrasonic Physics*. 1st edition, New York, American Elsevier Publishing Company.

Sapozhnikov, O. A. (2015). Chapter 2, Editor(s): Juan A. Gallego-Juárez, Karl F. Graff. *High-Intensity Ultrasonic Waves in Fluids: Nonlinear Propagation and Effects, Power, Ultrasonics*, Woodhead Publishing, 9–35. ISBN 978-1-84569-989-5.

Shanmugam, A., Chandrapala, J., & Ashokkumar, M. (2012). The effect of ultrasound on the physical and functional properties of skim milk. *Innovative Food Science and Emerging Technologies*, 16, 251–258.

Silva, M., & Chandrapala, J. (2021). Ultrasound in dairy processing. In: Editor(s): Kai Knoerzer, Kasiviswanathan Muthukumarappan. *Innovative Food Processing Technologies*, Elsevier, 439–464. ISBN 978-0-12-815782-4.

Sim, J. Y., Beckman, S. L., Anand, S., & Martínez-Monteagudo, S. I. (2021). Hydrodynamic cavitation coupled with thermal treatment for reducing counts of B. coagulans in skim milk concentrate. *Journal of Food Engineering*, 293, 110382.

Tan, M. C., Chin, N. L., Yusof, Y. A., & Abdullah, J. (2016). Effect of high power ultrasonic treatment on whey protein foaming quality. *International Journal of Food Science and Technology*, 51(3), 617–624.

Tan, M. C., Chin, N. L., Yusof, Y. A., Taip, F. S., & Abdullah, J. (2015). Characterisation of improved foam aeration and rheological properties of ultrasonically treated whey protein suspension. *International Dairy Journal*, 43, 7–14.

van Wijngaarden, L. (2016). Mechanics of collapsing cavitation bubbles. *Ultrasonics Sonochemistry*, 29, 524–527.

Villamiel, M., & de Jong, P. (2000). Influence of high-intensity ultrasound and heat treatment in continuous flow on fat, proteins, and native enzymes of milk. *Journal of Agricultural and Food Chemistry*, 48(2), 472–478.

Wang, D., & Sakakibara, M. (1997). Lactose hydrolysis and β-galactosidase activity in sonicated fermentation with Lactobacillus strains. *Ultrasonics Sonochemistry*, 4(3), 255–261.

Zhang, R., Pang, X., Lu, J., Liu, L., Zhang, S., & Lv, J. (2018). Effect of high intensity ultrasound pretreatment on functional and structural properties of micellar casein concentrates. *Ultrasonics Sonochemistry*, 47, 10–16.

Zisu, B., Lee, J., Chandrapala, J., Bhaskaracharya, R., Palmer, M., Kentish, S., & Ashokkumar, M. (2011). Effect of ultrasound on the physical and functional properties of reconstituted whey protein powders. *The Journal of Dairy Research*, 78(2), 226.

Zisu, B., Sciberras, M., Jayasena, V., Weeks, M., Palmer, M., & Dincer, T. D. (2014). Sonocrystallisation of lactose in concentrated whey. *Ultrasonics Sonochemistry*, 21(6), 2117–2121.

Chapter 7

Supercritical CO₂ Extraction Process

Kaavya Rathnakumar, Ahmed Hammam, Juan Camilo Osorio, and Sergio I. Martinez-Monteagudo

CONTENTS

7.1	Introduction	93
7.2	Theory on Supercritical CO₂ Extraction	94
7.3	Supercritical Process in Dairy Products	95
	7.3.1 Fluid Milk	95
	7.3.2 Butter	97
	7.3.3 Butter Oil	98
	7.3.4 Cheese	98
	7.3.5 Anhydrous Milk Fat	100
	7.3.6 Buttermilk Derivatives	100
	7.3.7 Beta-Serum and WPPC	101
7.4	Conclusions	101
References		101

7.1 INTRODUCTION

Supercritical fluids refer to heated and compressed fluids above their critical state – in a supercritical state. Fluids within the supercritical state exhibit liquid-like density and gas-like viscosity, which are desirable properties for extraction and mass transfer operations. Thus, supercritical fluids have been used to perform a number of industrial processes, including decaffeination of green coffee beans, production of hops extract, extraction of spices and herbs, extraction of essential oils, production of antioxidants, cleaning and decontamination of rice, impregnation of wood and polymers, and cleaning of cork (Eggers and Lack, 2012).

Over the past 40 years, the applications of supercritical fluids have been adopted in many operations. Nowadays, supercritical fluids are used to design particles for drug delivery, treatment of polymers, chemical and enzymatic reactions, extraction and recovery of valuable compounds, and treatment of wastewater. In the dairy industry, carbon dioxide (CO_2) has been the fluid of choice due to its costs, inflammability, and toxicity. Supercritical fluid carbon dioxide (SCO_2) has been used to perform a number of manufacturing operations. Microbial and enzymatic inactivation, carbonation of milk, fractionation of milk fat, and extraction of target compounds are examples of relevant applications of SCO_2. This chapter summarizes relevant lab-scale applications of SCO_2 within the dairy manufacturing framework.

DOI: 10.1201/9781003138716-7

7.2 THEORY ON SUPERCRITICAL CO₂ EXTRACTION

A given fluid that has been heated and pressurized above its critical point is considered as supercritical fluid. Such conditions result in the alteration of the phase behavior and transport properties of the system. Figure 7.1 illustrates the phase transition lines for melting, evaporation, and sublimation for a given fluid. The transition line corresponding to condensation and evaporation disappears after the critical point, where a continuous transfer from gas to liquid and liquid to gas occurs without a phase change (Eggers, 2012). Above the critical point, condensation and evaporation for a given fluid merge into a single phase. In this region, there is no clear distinction between gas and liquid, and it is known as the supercritical region.

The interfacial phenomena are reduced in supercritical fluids due to the negligible movement of particles across the contacting phases. This is an important consideration since the interfacial phenomena play a crucial role in heat and mass transfer operations, including extraction. Thermodynamically, the surface tension can be expressed as a function of temperature and pressure according to Equation 7.1.

$$\left(\frac{\partial \gamma}{\partial p}\right)_{T,A} = \left(\frac{\partial V}{\partial A}\right)_{p,T} \tag{7.1}$$

where γ = surface tension; p = pressure; V = volume of the system; A = surface area. A closer inspection of Equation 7.1 reveals that pressure will reduce the surface tension when molecular concentration occurs as a result of the applied pressure. On the other hand, an increase in surface tension will occur when the volume contraction of the continuous phase overcomes the attractive forces in the interfacial volume. Moreover, the supercritical region is characterized by the low-density region of gas and the high-density region of liquid, which enhances the extraction process. The speciality of these fluids is their different physicochemical properties such as density, diffusivity, viscosity, and dielectric constant. A number of fluids have been evaluated for use as supercritical fluids based on their critical temperature and pressure, costs, inflammability, and toxicity. Table 7.1 provides an overview of selected compounds used as supercritical fluids.

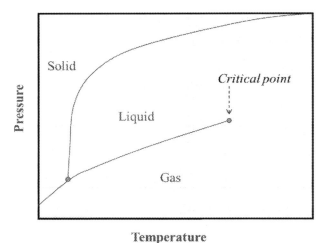

Figures 7.1 Illustration of the supercritical state for a given fluid.

TABLE 7.1 SELECTED FLUIDS AND THEIR SUPERCRITICAL CONDITIONS

Fluid	Generic formula	T_c (K)	P_c (MPa)
Carbon dioxide	CO_2	304.1	7.38
Water	H_2O	647.0	22.06
Methane	CH_4	190.4	4.60
Ethane	C_2H_6	305.3	4.87
Propane	C_3H_8	369.8	4.25
Ethylene	C_2H_2	282.4	5.04
Propylene	$(C_3H_6)_n$	364.9	4.60
Methanol	CH_4O	512.6	8.09
Ethanol	C_2H_5OH	513.9	6.14
Acetone	C_3H_6O	508.1	4.70

T_c – critical temperature; P_c – critical pressure

Due to their low viscosity and relatively high diffusivity, supercritical fluids have more enhanced transport properties than liquids, can diffuse easily through solid materials, and can, therefore, give faster extraction rates (Amaral et al., 2017; da Silve et al., 2016). One of the main characteristics of a supercritical fluid is the possibility of modifying the density of the fluid by changing its pressure and/or temperature. Since density is related to solubility, by altering the extraction pressure, the solvent strength of the fluid can be modified (Pouliot et al., 2014). Carbon dioxide (CO_2) is the most widely used solvent as it is harmless to human health and the environment, and its critical temperature (31.04° C) allows the preservation of extracted bioactive molecules and, volatile compounds, minimizing the physicochemical, nutritional, and sensorial characteristics of the food (Bezerra et al., 2020).

Overall, solids are processed batchwise or semi-continuously, while liquids are processed in a continuous mode. Figure 7.2 represents a typical supercritical extraction of solids. The process involves two steps – extraction and separation. In this extraction, the raw material is placed in an extractor vessel connected to temperature and pressure controllers for maintaining the desired conditions. Then the extractor vessel is pressurized with the fluid by a pump. Once the fluid and the dissolved compounds are transported to separators, where the products are collected through a tap located in the lower part of the separators, the system is decompressed and the supercritical fluid turns into gaseous. Finally, the fluid is regenerated and cycled or released to the environment (Gil-Chavez et al., 2013). The supercritical fluid extraction (SFE) is a rapidly developing method to obtain bioactive compounds under mild temperature conditions as most bioactives are thermally liable. CO_2 is considered to extract apolar and moderately polar compounds. However, when smaller volumes of more polar solvents (modifiers or cosolvents) such as water, ethyl acetate, ethanol, or methanol are added the polarity is modified and more polar compounds can be extracted due to the CO_2's increased solvency and the solubility of the polar compound.

7.3 SUPERCRITICAL PROCESS IN DAIRY PRODUCTS

7.3.1 Fluid Milk

Fluid milk has been treated with supercritical CO_2 for mainly four purposes: microbial and enzyme inactivation, pasteurization, and removal of cholesterol.

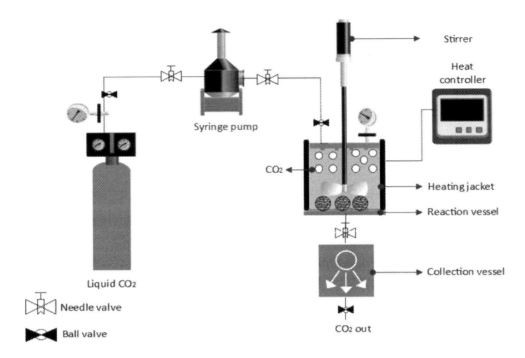

Figure 7.2 Schematic representation of Super-critical Fluid extraction.

i) *Microbial inactivation* – the mechanism of inactivation of SCO_2 is related to the alteration of the cell membrane, affecting microbial metabolism. This is because of the decarboxylation reaction which happens in the microbial metabolism caused due to the accumulation of pressure and addition of carbon-di-oxide causing lethal damages (Singh et al., 2018). SCO_2 helped in a reduction in the microbial load of *Escherichia coli* and aerobic mesophilic bacteria in human milk (Singh et al., 2018). Overall, the antimicrobial efficiency of SCO_2 depends on the temperature and pressure within the medium (8-15 MPa) and high-pressure range (>20 MPa). The lethality of SCO_2 has been reviewed elsewhere (Garcia-Gonzalez et al., 2007).

ii) *Enzyme inactivation* – alkaline phosphatase is an endogenous enzyme present in milk and it is considered to be more resistant to heat compare to most pathogenic bacteria. The employment of SCO_2 resulted in about 94% inactivation at a CO_2 to milk mass ratio of 0.05, 70°C, and 80 bar with a residence time of 30 min (Ceni et al., 2016). Enzymes are thought to be inactivated by the displacement of the oxygen and lowering of the pH, which causes the inactivation of enzymes for the regulation of the metabolic process (Amaral et al., 2017).

iii) *Milk pasteurization* – SCO_2 has been used to pasteurize skim milk. Di Giacomo et al. (2009) treated raw skim milk at 150 bar, 35-40°C and a CO_2 to milk feed ratio of 0.33. These authors reported a shelf-life of 35 days.

iv) *Removal of cholesterol* – SCO2 enhances the solvent power as a liquid and it also behaves as a gas with no surface tension. This selectivity has caused the separation efficiency of cholesterol and therefore it is a pressure and temperature-dependent phenomenon

(Bradley 1989). Another study used supercritical CO$_2$ (20.7 MPa/68°C) to reduce the cholesterol content in whole milk powder (Chitra et al., 2015), where about 55% of the total cholesterol was removed without any alteration in the free fatty acids, lightness value, and solubility index.

7.3.2 Butter

Butter is manufactured by separating the cream from milk, followed by ripening and churning the cream to produce the butter (Figure 7.3). Supercritical CO$_2$ has been utilized to extract and fractionate palm kernel oil (Zaidul et al., 2006) and cocoa butter (Saldaña et al., 2002). The fractionation of milk fat is utilized nowadays more to improve the functional characteristics compared to the nutritional value (Kontkanen et al., 2011).

Milk fat can be used in different applications based on the melting characteristics, such as ice-cream that requires high melting fat fractions and spreadable butter, bakery products, or reconstituted milk powder that requires low melting fat fractions. Unsaturated fatty acids and short chain saturated fatty acids are fractionated in the liquid olein portion; while flavor compounds, pigments, cholesterol, vitamin A, and long chain saturated fatty acids are separated in the solid stearin portion during the dry-melt fractionation process (Macías et al., 2014). Additionally, milk fat can increase the nutritional value of dairy products by decreasing the saturated fatty acids concentration. It has been found that the oxidative stability is increasing with the unsaturated fatty acids in the fat fractions that have low melting properties (El-Aziz, 2008). Linoleic acid has been concentrated from anhydrous milk fat using supercritical CO$_2$ extraction (Romero et al., 2000). These authors suggested that the fat fraction obtained from supercritical CO$_2$ extraction can increase the nutritional value as well as the functional characteristics of dairy products due to its high melting properties that related to several factors such as unsaturated fatty acids, the high level of conjugated linoleic acid, and β-carotene (Romero et al., 2000). Another study has utilized supercritical CO$_2$ extraction with different enzymes, namely Lipozyme RM IM, Novozyme 435, and immobilized enzyme from *Candida rugosa* to produce a high level of free fatty acids, such as conjugated linoleic acid from anhydrous milk fat (Prado et al., 2012). These authors evaluated different pressures (20 to 30 MPa) and temperatures (45 and 65 °C) with different fat to water ratios (1:5–1:30 mol/mol). The highest concentration of free fatty acids (>85.0% w/w) was reached at 23 MPa and a fat to: water ratio of 1:5 using Lipozyme TL IM at 55 °C. The highest

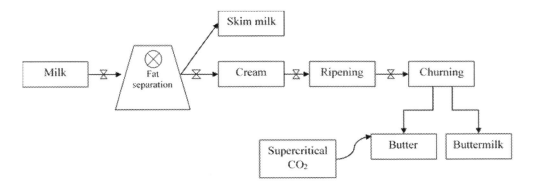

Figure 7.3 Schematic of manufacture of butter and processing using supercritical CO$_2$.

concentration (6.81 mg/g fat) of conjugated linoleic acid was reached at 30 MPa and a fat to: water ratio of 1:30 using Lipozyme TL IM at 55 °C (Prado et al., 2012).

7.3.3 Butter Oil

Butter oil and ghee are produced from cream or butter with differences in the method of manufacture (Figure 7.4). Chen et al. (1992a) fractionated butter oil into triglycerides with SCO_2 at different temperatures (35, 40, 50, and 60°C) and pressures (10.3, 13.8, 17.2, 20.7, 24.1, and 27.6 MPa). Overall, these authors reported that the extraction of triglycerides was more efficient at high pressures and low temperatures. Pressure within the low spectrum resulted in a higher concentration of short and medium fatty acids in the fractions of triglyceride, with a low concentration of long fatty acids (Chen et al., 1992a).

Fouad et al. (1994) fractionated butter oil with SCO_2 and reported that about ~13.8 MPa and 35°C increased the concentration of low molecular triglycerides (short and medium fatty acids). The obtained fraction showed improved spreadability compared with the untreated samples. Mohamed et al. (1998) reduced cholesterol and fractionated butter oil with SCO_2 using a flow rate of 1.4-2.0 g/min, 27.6 MPa, and 40°C. These authors were able to decrease the cholesterol content from 2.5 to 0.1 mg/g. Similarly, Fatouh et al. (2007) used multiple extractions of SCO_2 to fractionate butter oil produced from buffalo milk. The fractionation was performed by increasing the pressure from 10.9 to 15 MPa at 50°C and 22.3 to 40.1 MPa at 70°C. It was reported that low pressures resulted in a high concentration of cholesterol, saturated fatty acids, and short fatty acids with lower melting properties, while increasing the pressures led to elevating the content of long fatty acids as well as unsaturated fatty acids with higher melting properties. Supercritical CO_2 was also utilized to extract the volatile flavor components from ghee (Duhan et al., 2020), where several parameters were analyzed, such as pressure (3-8 MPa), time (19-221 min), and temperature (3-38 °C). The highest level of volatile compounds extracted from ghee was at 7 MPa and 10 °C for 60 min.

7.3.4 Cheese

Cheese is manufactured by fermenting and coagulating milk using starter cultures and rennet. After coagulation, cheese is cut, pressed, and then ripened (Figure 7.5). Modern consumers

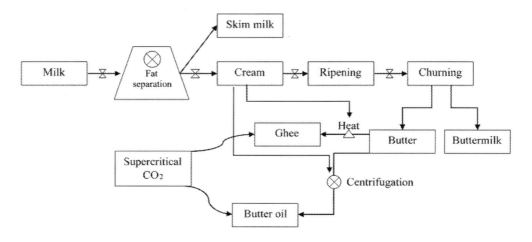

Figure 7.4 Schematic of manufacture of butter oil or ghee and processing using supercritical CO_2.

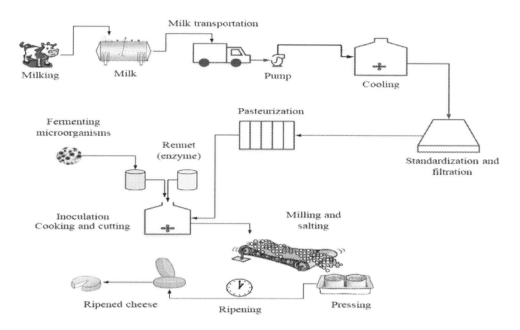

Figure 7.5 Schematic diagram of cheese manufacture.

prefer low-fat cheese. However, low-fat cheeses have different characteristics relative to full-fat cheese, including texture, sensory, and color properties which are not favorable to consumers. As a result, many researchers have tried several methods to improve the characteristics of low-fat cheese, such as using different starter cultures (e. g., probiotic and exopolysaccharide producing types), additives, procedures, and fat replacers (Sanchez-Macías et al., 2013).

Supercritical fluid extraction (SFE) has been utilized to remove the fat from cheddar and parmesan (Yee et al., 2007). These authors found that the efficiency of fat removal from cheddar cheese was high at 20 MPa and 35 °C. In the case of parmesan, the was removed at 35 MPa and 40 °C with 1000 g/kg of CO_2. It has been reported that using SCO_2 (1000 g of CO_2 at 35 °C) and pressure (10, 20, 30, or 40 MPa) resulted in more fat reduction in artisanal goat cheese (Majorero) as compared to Gouda goat cheese (Sanchez-Macías et al., 2013). Interestingly, the moisture content is significantly reduced in cheese is treated with SCO_2 (Sanchez-Macías et al., 2013; Yee et al., 2007, 2008). The loss of moisture depends on the pressure as well as the exposure time. Additionally, cheeses treated with SCO_2 exhibited flavors that are not typically found in low-fat milk (Yee et al., 2007). It has been found that nonpolar lipids, including triacylglycerides, free fatty acids, and cholesterol were removed from the SCO_2-treated cheese, while polar lipids, such as phospholipids existed in the cheese matrix (Sanchez-Macías et al., 2013; Yee et al., 2007).

The application of SCO_2 in cheeses reduced the microbial load, extending the shelf-life. Sanchez-Macías et al. (2013) studied the effect of SCO_2 (10 MPa and 35°C for 50 min) on the total aerobic bacterial count, lactococci count, and lactobacilli count in Majorero cheese and goat Gouda cheese. A reduction of about 1.5-2.5 log cfu/g was found regardless of the type of cheese. Perrut et al. (2012) concluded that the reduction in the microbial count is more affected by the cheese matrix when the pressure is > 20 MPa. Overall, SCO_2 technology can be used to reduce the fat content, microbial count, and nonpolar lipids in cheese; and this depends on the cheese type as well as the parameters.

7.3.5 Anhydrous Milk Fat

Fractionation of milk fat is employed to improve the functionality and nutritional properties of milk fat. Fractionation of anhydrous milk fat (AMF) has been carried out since the 1980's, where a pressure range of 100-350 bar and temperatures between 50 and 70° C separated long and short-chain fatty acids (Arul et al., 1987). Subsequent studies on fractionation under continuous SCO_2 resulted in the separation of five fractions at a pressure range between 24.1 - 3.4 MPa at 40-75 °C (Bhaskar et al., 1993). A multi-step sequential extraction of SCO_2 using a simultaneous pressure of 10.9 and 15 MPa at 50 °C and 22.3 and 40.1 MPa at 70 °C showed the ability to fractionate buffalo butter oil (Fatouh et al., 2007). At low pressure levels, a fraction containing cholesterol, short-chain and saturated fatty acids was obtained. On the contrary, higher pressure levels enriched the concentration of long-chain and unsaturated fatty acids. Buyukbese et al. (2017) obtained six fractions from milk fat with SCO_2 under the pressure condition 10-36 MPa, and temperature 40-60 °C. The obtained fractions were analyzed in terms of thermal behavior, iodine value, and color index. It was also observed that fractions produced at low pressures had lower melting behavior than those obtained at high pressures (Buyukbese et al., 2017). Huber et al. (1996) removed 97% of the cholesterol from milk fat with SCO_2 (8-40 MPa and 40-70 °C). SCO_2 has also been used to increase the concentration of conjugated linoleic acid (CLA), a bioactive lipid naturally present in milk fat (Chen & Park, 2019). Romero et al. (2000) concentrated CLA from anhydrous milk fat with SCO_2. Prado et al. (2012) concentrated CLA from milk fat using SCO_2 and enzymatic hydrolysis and the maximum production of free fatty acids (FFAs) was 98% from triglycerides.

7.3.6 Buttermilk Derivatives

Buttermilk is a by-product obtained from the churning of butter which has a rich amount of milk fat globular membrane (MFGM). Figure 7.6 illustrates a process combining membrane filtration with SCO_2. The membranes are surrounded by the milk fat globules, such arrangement has caused the milk fat to be emulsified and dispersed within the milk. MFGM contains proteins, minerals, neutral lipids, and enzymes (Danthine et al., 2000). The most significant is the phospholipids, which account for about 4.49 g per 100 g of fat, which is 6.5 times more than the phospholipids

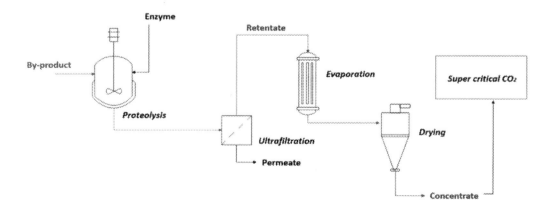

Figure 7.6 Illustration of SCO_2 extraction of phospholipids in combination with membrane filtration.

found in milk (Avalli & Contarini, 2005). Phospholipids have gained a potential interest in recent times due to their health and nutrition benefits (Ortega-Anaya & Jimenez-Flores, 2019).

The major challenge in the isolation and concentration of the MFGM is the presence of skim milk solids, mostly the casein micelles which inhibit the concentration process. Astaire et al. (2003) developed a two-step process to concentrate milk phospholipids up to 19% of the total lipids. The process consisted of microfiltration followed by supercritical CO_2 extraction. Similarly, Costa et al. (2010) concentrated buttermilk using the two-step process following by spray drying, obtaining a powder of up to 61% of phospholipids. Another investigation on the use of enzymatic treatment followed by ultrafiltration showed an 8.5-fold increase in the milk phospholipids. Moreover, these authors further concentrated the buttermilk with supercritical CO_2 to obtained a product containing up to 56% milk phospholipids. Sprick et al. (2019) removed neutral lipids from whey protein phospholipid concentrate by supercritical CO_2, while the fraction containing the polar lipids was further concentrated (up to 26% of MPLs) using ethanol (10-20%) as a co-solvent during supercritical CO_2. Ubeyitogullari and Rizvi (2020) reported an MPLs fraction containing 76% using sequential supercritical carbon dioxide and ethanol supercritical carbon dioxide.

7.3.7 Beta-Serum and WPPC

Beta-serum (BS) is a by-product obtained from the phase inversion of anhydrous milk fat containing approximately 60% fat. Researchers have successfully extracted nonpolar and polar lipids from BS. Catchpole et al. (2008) used dimethyl ether (DME) and CO_2 to extract both neutral and polar lipids. Spray-dried beta-serum first extracted a majority of neutral lipids by SFE with neat CO2 under 300 bar and 40°C. The PLs enriched residual was then re-extracted by SFE with DME under 40 bar and 60°C. The polar lipid extracts were analyzed and contained 70% PLs. Whey protein phospholipid concentrate (WPPC) also known as pro-cream is a by-product produced from the processing of cheese which is obtained during the micro-filtration of whey protein isolate (Levin et al., 2016). Sprick et al. (2019) developed a two-step process with SCO_2 and ethanol as a co-solvent to obtain a fraction rich in phospholipids (26.26 g of total PL/100g of fat at 35.0 MPa, 40°C, and 15% ethanol).

7.4 CONCLUSIONS

Supercritical CO_2 has been used to perform a number of manufacturing operations within the dairy industry, including extraction, fractionation, and pasteurization. This technology has the potential to be established as an essential unit operation in the near future. Limited research efforts have been conducted to generate knowledge of the phase equilibrium and the mass transfer kinetics, which is essential for the specification of supercritical processes.

REFERENCES

Amaral, G. V., Silva, E. K., Cavalcanti, R. N., Cappato, L. P., Guimaraes, J. T., Alvarenga, V. O., Esmerino, E. A., Portela, J. B., Sant' Ana, A. S., Freitas, M. Q., Silva, M. C., Raices, R. S. L., Meireles, M. A. A., & Cruz, A. G. (2017). Dairy processing using supercritical carbon dioxide technology: Theoretical fundamentals, quality and safety aspects. *Trends in Food Science & Technology, 64*, 94–101.

Arul, J., Boudreau, A., Makhlouf, J., Tardif, R., & Sahasrabudhe, M. R. (1987). Fractionation of anhydrous milk fat by superficial carbon dioxide. *Journal of Food Science*, *52*(5), 1231–1236.

Astaire, J. C., Ward, R., German, J. B., & Jiménez-Flores, R. (2003). Concentration of polar MFGM lipids from buttermilk by microfiltration and supercritical fluid extraction. *Journal of Dairy Science*, *86*(7), 2297–2307.

Avalli, A., & Contarini, G. (2005). Determination of phospholipids in dairy products by SPE/HPLC/ELSD. *Journal of Chromatography. A*, *1071*(1–2), 185–190.

Bezerra, F. W. F., de Oliveira, M. S., Bezerra, P. N., Cunha, V. M. B., Silva, M. P., da Costa, W. A., Pinto, R. H. H., Cordeiro, R. M., da Cruz, J. N., Chaves Neto, A. M. J., & Carvalho, J. R. N. (2020). Extraction of bioactive compounds. In: Inamuddin, A. M. Asiri, & A. M. B. T.-G. S. P. for C. and E. E. and S. Isloor (Eds.), *Green Sustainable Process for Chemical and Environmental Engineering and Science* (pp. 149–167). Elsevier.

Bhaskar, A. R., Rizvi, S. S. H., & Sherbon, J. W. (1993). Anhydrous milk fat fractionation with continuous countercurrent supercritical carbon dioxide. *Journal of Food Science*, *58*(4), 748–752.

Bradley, R. L. (1989). Removal of cholesterol from milk fat using supercritical carbon dioxide. *Journal of Dairy Science*, *72*(10), 2834–2840.

Buyukbese, D., Rousseau, D., & Kaya, A. (2017). Composition and shear crystallization of milkfat fractions extracted with supercritical carbon dioxide. *International Journal of Food Properties*, *20*(sup3), S3015–S3026.

Catchpole, O. J., Tallon, S. J., Grey, J. B., Fletcher, K., & Fletcher, A. J. (2008). Extraction of lipids from a specialist dairy stream. *The Journal of Supercritical Fluids*, *45*(3), 314–321.

Ceni, G., Fernandes Silva, M., Valério Jr., C., Cansian, R. L., Oliveira, J. V., Dalla Rosa, C., & Mazutti, M. A. (2016). Continuous inactivation of alkaline phosphatase and Escherichia coli in milk using compressed carbon dioxide as inactivating agent. *Journal of CO2 Utilization*, *13*, 24–28.

Chen, H., Schwartz, S. J., & Spanos, G. A. (1992a). Fractionation of butter oil by supercritical carbon dioxide. *Journal of Dairy Science*, *75*(10), 2659–2669.

Chen, P. B., & Park, Y. (2019). Conjugated linoleic acid in human health: Effects on weight control. In: R. R. B. T.-N. in the P. and T. of A. O. (Second E. Watson) (Ed.), *Nutrition in the Prevention and Treatment of Abdominal Obesity* (pp. 355–382). Elsevier.

Chitra, J., Deb, S., & Mishra, H. N. (2015). Selective fractionation of cholesterol from whole milk powder: Optimisation of supercritical process conditions. *International Journal of Food Science and Technology*, *50*(11), 2467–2474.

Costa, M. R., Elias-Argote, X. E., Jiménez-Flores, R., & Gigante, M. L. (2010). Use of ultrafiltration and supercritical fluid extraction to obtain a whey buttermilk powder enriched in milk fat globule membrane phospholipids. *International Dairy Journal*, *20*(9), 598–602.

da Silva, R. P. F. F., Rocha-Santos, T. A. P., & Duarte, A. C. (2016). Supercritical fluid extraction of bioactive compounds. *TrAC Trends in Analytical Chemistry*, *76*, 40–51.

Danthine, S., Blecker, C., Paquot, M., Innocente, N., & Deroanne, C. (2000). Progress in milk fat globule membrane research: A review. *Lait*, *80*(2), 209–222.

Di Giacomo, G., Taglieri, L., & Carozza, P. (2009). Pasteurization and sterilization of milk by supercritical carbon dioxide treatment. *Proceedings of the 9th International Symposium on Supercritical Fluids* (pp. 18–20).

Duhan, N., Sahu, J. K., & Naik, S. N. (2020). Sub-critical CO2 extraction of volatile flavour compounds from ghee and optimization of process parameters using response surface methodology. *LWT*, *118*, 108731.

El-Aziz, M. A. (2008). Properties of butter oil fractions and its formulated emulsions. *Egyptian Journal of Dairy Science*, *36*(1), 53.

Eggers, R. (2012). Basic engineering aspects. In: Eggers, R. (Ed.), *Industrial High Pressure Applications, Processes, Equipment and Safety*, 1st edn (pp. 7–48). Weinheim, Wiley-VCH Verlag GmbH & Co.

Eggers, R., & Lack, E. (2012). Supercritical processes. In: Eggers, R. (Ed.), *Industrial High Pressure Applications, Processes, Equipment and Safety*, 1st edn (pp. 169–209). Weinheim, Wiley-VCH Verlag GmbH & Co.

Fatouh, A. E., Mahran, G. A., El-Ghandour, M. A., & Singh, R. K. (2007). Fractionation of buffalo butter oil by supercritical carbon dioxide. *LWT - Food Science and Technology*, *40*(10), 1687–1693.

Fouad, F. M., Farrell, P. G., Voort, F. R., & Marshall, W. D. (1994). Modification of butter oil: Effects of supercritical carbon dioxide extraction and intereseterification. *Journal of Food Lipids*, *1*(3), 195–219.

References

Garcia-Gonzalez, L., Geeraerd, A. H., Spilimbergo, S., Elst, K., Van Ginneken, L., Debevere, J., Van Impe, J. F., & Devlieghere, F. (2007). High pressure carbon dioxide inactivation of microorganisms in foods: The past, the present and the future. *International Journal of Food Microbiology, 117*(1), 1–28.

Huber, W., Molero, A., Pereyra, C., & Martinez de la Ossa, E. (1996). Dynamic supercritical CO2 extraction for removal of cholesterol from anhydrous milk fat. *International Journal of Food Science and Technology, 31*(2), 143–151.

Joana Gil-Chávez, G., Villa, J. A., Fernando Ayala-Zavala, J., Basilio Heredia, J., Sepulveda, D., Yahia, E. M., & González-Aguilar, G. A. (2013). Technologies for extraction and production of bioactive compounds to be used as nutraceuticals and food ingredients: An overview. *Comprehensive Reviews in Food Science and Food Safety, 12*(1), 5–23.

Kontkanen, H., Rokka, S., Kemppinen, A., Miettinen, H., Hellström, J., Kruus, K., Marnila, P., Alatossava, T., & Korhonen, H. (2011). Enzymatic and physical modification of milk fat: A review. *International Dairy Journal, 21*(1), 3–13.

Levin, M. A., Burrington, K. J., & Hartel, R. W. (2016). Whey protein phospholipid concentrate and delactosed permeate: Applications in caramel, ice cream, and cake. *Journal of Dairy Science, 99*(9), 6948–6960.

Macías, D. S., Castro, N., Argüello, A., & Flores, R. J. (2014). Supercritical fluid extraction application on dairy products and by-products. In: Osborne J. (Ed.), *Handbook on Supercritical Fluids: Fundamentals, Properties and Applications*, Nova Science Publishers, Inc (pp. 281–300).

Mohamed, R. S., Neves, G. B. M., & Kieckbusch, T. G. (1998). Reduction in cholesterol and fractionation of butter oil using supercritical CO_2 with adsorption on alumina. *International Journal of Food Science and Technology, 33*(5), 445–454.

Ortega-Anaya, J., & Jiménez-Flores, R. (2019). Symposium review: The relevance of bovine milk phospholipids in human nutrition—Evidence of the effect on infant gut and brain development. Journal of Dairy Science, *102*(3), 2738–2748.

Perrut, M. (2012). Sterilization and virus inactivation by supercritical fluids (a review). *The Journal of Supercritical Fluids, 66*, 359–371.

Prado, G. H. C., Khan, M., Saldaña, M. D. A., & Temelli, F. (2012). Enzymatic hydrolysis of conjugated linoleic acid-enriched anhydrous milk fat in supercritical carbon dioxide. *The Journal of Supercritical Fluids, 66*, 198–206.

Pouliot, Y., Conway, V., & Leclerc, P.-L. (2014). Separation and concentration technologies in food processing. In: Clark, S., S. Jung, & B. Lamsal (Eds.), *Food Processing* (pp. 33–60). John Wiley &amp Sons, Ltd.

Romero, P. K., Rizvi, S. S. H., Kelly, M. L., & Bauman, D. E. (2000). Short communication: Concentration of conjugated linoleic acid from milk fat with a continuous supercritical fluid processing system. *Journal of Dairy Science, 83*(1), 20–22.

Saldaña, M. D. A., Mohamed, R. S., & Mazzafera, P. (2002). Extraction of cocoa butter from Brazilian cocoa beans using supercritical CO2 and ethane. *Fluid Phase Equilibria, 194–197*, 885–894.

Sánchez-Macías, D., Laubscher, A., Castro, N., Argüello, A., & Jiménez-Flores, R. (2013). Effects of supercritical fluid extraction pressure on chemical composition, microbial population, polar lipid profile, and microstructure of goat cheese. *Journal of Dairy Science, 96*(3), 1325–1334.

Singh, S. K., Pavan, M. S., N, S. P., & Kant, R. (2018). Applications of super critical fluid extraction in milk and dairy industry: A review. *Journal of Food Processing and Technology, 9*(12), 1–8.

Sprick, B., Linghu, Z., Amamcharla, J. K., Metzger, L. E., & Smith, J. S. (2019). Selective extraction of phospholipids from whey protein phospholipid concentrate using supercritical carbon dioxide and ethanol as a co-solvent. *Journal of Dairy Science, 102*(12), 10855–10866.

Ubeyitogullari, A., & Rizvi, S. S. H. (2020). Production of high-purity phospholipid concentrate from buttermilk powder using ethanol-modified supercritical carbon dioxide. *Journal of Dairy Science, 103*(10), 8796–8807.

Yee, J. L., Khalil, H., & Jiménez-Flores, R. (2007). Flavor partition and fat reduction in cheese by supercritical fluid extraction: Processing variables. *Le Lait, 87*(4–5), 269–285.

Yee, J. L., Walker, J., Khalil, H., & Jiménez-Flores, R. (2008). Effect of variety and maturation of cheese on supercritical fluid extraction efficiency. *Journal of Agricultural and Food Chemistry, 56*(13), 5153–5157.

Zaidul, I. S. M., Norulaini, N. A. N., Omar, A. K. M., & Smith, R. L. (2006). Supercritical carbon dioxide (SC-CO2) extraction and fractionation of palm kernel oil from palm kernel as cocoa butter replacers blend. *Journal of Food Engineering, 73*(3), 210–216.

Chapter 8

Application of Radiation-Based Processing of Dairy Products

M. Selvamuthukumaran and Sajid Maqsood

CONTENTS

8.1	Introduction	105
8.2	Ionizing Radiation Sources	106
8.3	Food Safety Enhancement and Consumer Perception	107
8.4	Effect of Irradiation on Organoleptic and Physicochemical Quality Characteristics	108
8.5	Application of Irradiation Techniques in the Dairy Sector	108
8.6	Conclusion	109
References		109

8.1 INTRODUCTION

Irradiation is a non-thermal processing technique that can be safely adopted for enhancing the stability of products. Several countries have approved the use of irradiated foods on a limited or unconditional basis. These techniques can be viable for reducing microbial spoilage and also increasing safety aspects even at low doses from the consumer point of view. There are wide varieties of irradiated foods available, but irradiation's application in the dairy sector is very limited. Therefore, there is ample scope for this technique to be applied in the dairy processing and preservation area.

Prepacked or packed foods either in cans or tin containers in the case of dried or frozen foods were exposed to ionizing radiations, i.e. X-rays or gamma rays, in a cabin for a specific time period, which can nullify the microbial spoilage. Food irradiation applications can be classified based on the dose of exposure as given in Table 8.1 and Figure 8.1.

Gamma rays are high-energy photons and they are produced by the spontaneous disintegration of radionuclides like Cobalt-60 and Cesium-137. A major characteristic of gamma rays is their high penetrating power, which facilitates their use in the treatment of bulk items. Since gamma rays do not give rise to neutrons, irradiated foods and their material are not made radioactive (Diehl, 1995). Cobalt-60 is the most commonly used radionuclide for food. Cobalt-60 emits ionizing radiation in the form of gamma rays and its advantages include deep penetration, uniformity of dose, etc.

TABLE 8.1 CLASSIFICATION OF IRRADIATION APPLICATION FOR FOOD USE

Nature of Dose	Quantity (Kilogray)	Purpose
Low	1	Sprouting inhibition of potatoes, grains, fruit ripening delay, insect disinfestation during grains storage
Medium	1–10	Enhancement of food stability, microbial load reduction in most of the foods, technological properties improvement
High	10–50	Virus elimination, achieving commercial sterilization

Note: modified from Loaharanu, 1985.

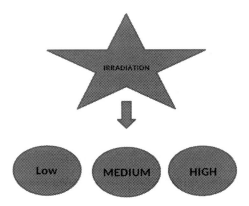

Figure 8.1 Classification of irradiation for food use.

8.2 IONIZING RADIATION SOURCES

There are three different kinds of ionizing radiation that are used for food processing, which include accelerated electrons, X-rays and gamma rays. Gamma rays won't give rise to neutrons and therefore the irradiated foods and their material won't be made radioactive (Diehl, 1995). Hirneisen et al. (2010) reported that the limitations of Cobalt-60 are its shorter half-life period of around 5.3 years and its slow treatment process for foods when compared to other sources of radiation.

Cesium-137 is not commercially used due to handling issues (Lee, 2004), even though it has a half-life period of 30 years. One has to follow the protocol of keeping the radioactive material at the top of an elevator that can either be moved down or up under the water, particularly when it is not in use. Place the material that wants to receive an irradiation dose in and around the radioactive material at the desired distance for receiving the essential or required treatment dose. Isotopes always emit radiation in various directions; therefore they can't be shut off when not in use (Moreira, 2006).

The second ionizing radiation source is accelerated electrons, which can be generated by electron-accelerating machines, which are known as electron accelerators. Microbial destruction is achieved by means of increasing the light speed; the electron beam accelerator will send the higher-energy electrons, which can ultimately inactivate the microorganisms, onto the product. The electron accelerators can be easily shut off when not in use, thus avoiding hazards caused by radioactive isotope usage. Tahergorabi et al. (2012) reported that electron beam processing will never change the temperature of foods to be processed and it also even allows application at very higher doses, i.e. 1,000 to even 100,000 kGy/s, as compared to gamma radiation, i.e. 0.01 to 1 Gy/s. The penetration depth seems to be only 8 to 10 cm for food products; therefore before exposing the foods to irradiation, the size of the food needs to be considered (Jaczynsky and Park, 2003).

For grains, the electron beam is found to be highly successful, and it can also be used for food preparation surface decontamination and as well as for grounded spices.

The third source is X-rays, which can be created from high-energy electron bombardment on a metal target. They have the advantage of higher penetration power as well as switch-off capability (Lee, 2004). An X-ray machine will have a maximum energy of 5 MeV, which is very much less than what can induce food radioactivity. The X-radiation can penetrate food shallowly when compared to gamma radiation but more deeply than the electron beam (Tahergorabi et al., 2012; Sadler et al., 2001).

Moreira (2006) prescribed that the efficiency of irradiation will depend upon the characteristics of the accelerator, the irradiation techniques used and other factors, which include shape, geometric dimensions, type of material used, etc. The irradiation dose is often measured in terms of a gray (Gy) where one Gy is equivalent to one energy joule that is being absorbed by one kilogram of substance. Therefore the dose rate is expressed in kGy/s. The absorbed total dose of the irradiated material is related to both dose and irradiation time. The dose will be decided on the basis of the nature of the food product as well as the purpose of radiation (Hirneisen et al., 2010).

8.3 FOOD SAFETY ENHANCEMENT AND CONSUMER PERCEPTION

Irradiation's effects on microorganisms depend upon the free radical reaction with DNA and microbe RNA. DNA possesses the largest molecular structure in a cell, therefore it exhibits a greater degree of sensitivity when compared to the ionizing radiation effects (Scott and Suresh, 2004). Free radicals will destroy the hydrogen bonds between the DNA molecules (double-stranded), which can avoid or stop replication, thereby eliciting cell death, which can exert a lesser effect over non-living tissues (IFST, 2006). The microbe resistance to radiation may vary depending upon the strain and also the physiological condition of strains used. Verma and Singh (2001) reported that cells that are under stress may project higher resistance to irradiation levels. The starved cells will exhibit higher D10 values, i.e. the radiation dose required to achieve 1 log reduction, i.e. 90% (Mendonca et al., 2004). The efficiency of irradiation will be affected by factors like temperature, irradiation dose and exposure time, the microbe's physiological state and the atmosphere surrounding the foods. The spore-forming bacteria are quite resistant and they will only result in a 2–3 log reduction of spores when less than 10 kGy is being used; this is also not sufficient to manufacture shelf-stable foods (Patterson, 2005). Viruses and fungi are more resistant to radiation effects when compared to bacteria. The D10 values are in the range of 1–3 kGy for fungi and even more for viruses (Niemira and Deschenes, 2005; Yu et al., 1996).

The use of food irradiation dates back more than 100 years and its specific food applications have been permitted and approved by more than 55 countries' national legislations. But still, the use of this technique is quite limited due to various environmental and as well as health and safety concerns aspects. The treatment of foods with these ionizing radiations may use either gamma or X-rays or electron beam, which are accepted in Europe as well as in the Asia Pacific region, in order to meet the international trade requirements (Kume et al., 2009; Luckman, 2002). Luckman (2002) reported that in several countries, this irradiation technique is used for minimizing pathogenic microbes, especially from a consumer health point of view and also as part of GMP (good manufacturing practice) and HACCP (Hazard Analysis Critical Control Point) initiatives. In the USA such foods can be marketed as irradiated foods after labeling either as "treated by irradiation" or with a TrDURA logo at the sales point (Ehlermann, 2009). The European Union will allow the optional use of these labels.

Many market surveys have been carried out to evaluate the consumer attitude towards irradiated foods (Resurreccion et al., 1995; Fox, 2002; Wilcock et al., 2004). Many consumers think that irradiated foods are radioactive and that they can pose a threat to consumers, including current and future generations (Slovic et al., 1995). Consumer preference for buying irradiated foods lies only with the proper food labeling (Enneking, 2004; Bond et al., 2008). The technology will be accepted if the processing information is provided in the form of labeling. US people have accepted it highly, but it is slow in the European Union, including the UK (Olsen, 1999).

8.4 EFFECT OF IRRADIATION ON ORGANOLEPTIC AND PHYSICOCHEMICAL QUALITY CHARACTERISTICS

Irradiation, a kind of non-thermal technique, can preserve the foods with lesser interruption of organoleptic properties of foods even at lesser doses. Kim et al. (2006) found that higher application of irradiation doses can lead to the deterioration of organoleptic characteristics of foods, especially flavor, taste, color and texture.

The treatment of products with higher doses of irradiation may lead to mild lipid decomposition (Nawar, 1983). The protein level may vary depending upon the dose of irradiation treatment (Stewart, 2001) and with respect to carbohydrates, they can undergo major degradation as reported by Bhat and Karim (2009).

8.5 APPLICATION OF IRRADIATION TECHNIQUES IN THE DAIRY SECTOR

The application of irradiation for dairy food products is very scanty due to its safety issues and consumer awareness, but the technique can be adopted based on consumer preference and safety concern awareness creation so that the technology can be implemented in the dairy processing sector.

Mohamed et al. (2020) used an innovative technique to increase probiotic bacterial activities as well as fermented milk quality by treating the strain *Lactobacillus casei* with a red laser at the wavelength of 632.7 nm before starting the fermentation process of skim milk.

The milk was irradiated for a period of 40 min with an irradiation dose capacity of 12 j/cm^2; the results show that the total antioxidative capacity of pretreated milk was enhanced compared to the control. The recorded antioxidant capacity for treated samples was 56% when compared to the control, which exhibited only 21%. The fermented milk with laser exposure recorded significant enhancement of DPPH scavenging as well as total antioxidant capacity when compared to fermented milk without any laser treatment. The irradiation-treated fermented milk sample with *L. casei* displayed excellent lactose fermentation capability as a result of the passing red laser with a recorded β-Galactosidase activity of 0.37 unit/mL compared to the control sample.

The antibacterial activity of strain *L. casei* is significantly improved as a result of laser dose treatment at 12 j/cm^2 when compared to the control or untreated sample. The red laser light passed through the fermented milk sample, which then showed increased assimilation levels of cholesterol, around 41% for the sample that received a laser dose of 12 j/cm² when compared to the untreated sample. It also enhanced the proteolytic activity of *L. casei* after exposure to different doses of laser treatment before the fermentation period. The maximum proteolytic activity of 6.3% was recorded for laser-exposed bacterial group samples for a period of 40 min when compared to the control sample, which recorded a proteolytic activity of 5.1%. Therefore they concluded that laser treatment of *L. casei* at 12 J/cm² before the skimmed milk fermentation process exhibited significant enhancement of antioxidant activity, proteolytic, antimicrobial and β-galactosidase activities. It also further decreased cholesterol as well as fermented skimmed milk lactose levels, thereby enhancing the skimmed milk fermentation process with the *L. casei* strain.

8.6 CONCLUSION

There is a huge scope to adopt this kind of non-thermal processing technique for processing and preservation of dairy products without affecting their quality by creating consumer awareness about irradiated foods, their efficacy and safety aspects so that such irradiated foods will reach all sorts of consumers in the near future without any hurdle.

REFERENCES

Bhat, R., Karim, A.A. 2009. Impact of radiation processing on starch. *Comprehensive Reviews in Food Science and Food Safety*, 8(2), 44–58.

Bond, C.A., Thilmany, D.D., Bond, J.K. 2008. What to choose? The value of label claims to fresh produce consumers. *Journal of Agricultural and Resources and Economics*, 33, 402–427.

Diehl, J.F. 1995. *Safety of Irradiated Foods*, 2nd ed. New York, Marcel Dekker.

Ehlermann, D.A. 2009. The RADURA-terminology and food irradiation. *Food Control*, 20(5), 526–528.

Enneking, U. 2004. Willingness-to-pay for safety improvements in the German meat sector: The case of the Q&S label. *European Review of Agriculture Economics*, 31(2), 205–223.

Fox, J.A. 2002. Influences on purchase of irradiated foods. *Food Technology*, 56, 34–37.

Hirneisen, K.A., Black, E.P., Cascarino, J.L., Fino, V.R., Hoover, D.G., Kniel, K.E. 2010. Viral inactivation in foods: A review of traditional and novel food processing technologies. *Comprehensive Reviews in Food Science and Food Safety*, 9(1), 3–20.

IFST 2006. The use of irradiation for food quality and safety. Institute of Food Science and Technology. http://www.ifst.org/document.aspx?id=122.

Jaczynski, J., Park, J.W. 2003. Microbial inactivation and electron penetration in Surimi seafood during electron beam processing. *Journal of Food Science*, 68(5), 1788–1792.

Kim, M.J., Park, J.G., Kim, J.H., Park, J.N., Lee, H.J., Kim, W.G., Lee, J.W., Byun, M.W. 2006. Combined effect of heat treatment and gamma irradiation on the shelf- stability and quality of packaged Kimchi during accelerated storage condition. *Korean Journal of Food Preservation*, 13, 531–537.

Kume, T., Furuta, M., Todoriki, S., Uenoyama, N., Kobayashi, Y. 2009. Status of food irradiation in the world. *Radiation Physics and Chemistry*, 78(3), 222–226.

Lee, S.Y. 2004. Irradiation as a method for decontaminating food. *Internet J Food Saf*, 3, 32–35.

Loaharanu, P. 1985. Food Irradiation-A viable technology for reducing postharvest losses of foods. *ASEAN Journal of Science and Technology for Development*, 2(1), 80–87.

Luckman, G.J. 2002. Food irradiation: Regulatory aspects in the Asia and Pacific region. *Radiation Physics and Chemistry*, 63(3–6), 285–288.

Mendonca, A.F., Romero, M.G., Lihono, M.A., Nannapaneni, R., Johnson, M.G. 2004. Radiation resistance and virulence of Listeria monocytogenes Scott A following starvation in physiological saline. *Journal of Food Protection*, 67(3), 470–474.

Mohamed, M.S.M., Elshaghabee, F.M.F., Alharbi, S.A., El-Hussein, A. 2020. The Prospective Beneficial Effects of Red Laser Exposure on Lactocaseibacillus casei Fermentation of Skim Milk. *Biology*, 9(9), 256; doi:10.3390/biology9090256.

Moreira, R.G. 2006. Food irradiation using electron-beam accelerators. In: Hui Y.H. (eds) *Handbook of Food Science, Technology and Engineering*, Vol 3. Boca Raton, FL, CRC Press. pp 1241–1249.

Nawar, W.W. 1983. Comparison of chemical consequences of heat and irradiation treatment of lipids. In: Elias P.S. and Cohen A.J. (eds) *Recent Advances in Food Irradiation*. Amsterdam, The Netherlands, Elsevier Biomedical Press. pp 115–127.

Niemira, B.A., Deschenes, L. 2005. Ionizing radiation processing of fruits and fruit products. In: Barrett D.M., Somogyi L. and Ramaswamy H. (eds) *Processing Fruits: Science and Technology*, 2nd ed. Boca Raton, FL, CRC Press. pp 221–260.

Olsen, S.O. 1999. Strength and conflicting valence in the measurement of food attitude and preferences. *Food Quality and Preference*, 10(6), 483–494.

Patterson, M. 2005. Food irradiation: Microbiological safety and disinfestations international symposium new frontier of irradiated food and non-food products. Bangkok, Thailand.

Resurreccion, A.V.A., Galvez, F.C.F., Fletcher, S.M., Misra, S.K. 1995. Consumer attitudes toward irradiated food: results of a new study. *Journal of Food Protection*, 58(2), 193–196.

Sadler, G., Chappas, W., Pierce, D.E. 2001. Evaluation of e-beam, gamma- and X-ray treatment on the chemistry and safety of polymers used with pre-packaged irradiated foods: A review. *Food Additives and Contaminants, Part A*, 18(6), 475–501.

Scott, S.J., Suresh, P. 2004. Irradiation and food safety. *Food Technology*, 58, 48–55.

Slovic, P., Malmfors, T., Krewski, D., Mertz, C.K., Neil, N., Bartlett, S. 1995. Intuitive toxicology. II. Expert and lay judgments of chemical risks in Canada. *Risk Analysis*, 15(6), 661–675.

Stewart, E.M. 2001. Food irradiation chemistry. In: Molins R.A. (eds) *Food Irradiation: Principles and Applications*. New York, USA, John Wiley & Sons, Inc. pp 37–76.

Tahergorabi, R., Matak, K.E., Jaczynski, J. 2012. Application of electron beam to inactivate Salmonella in food: Recent developments. *Food Research International*, 45(2), 685–694.

Verma, N.C., Singh, R.K. 2001. Stress-inducible DNA repair in Saccharomyces cerevisiae. *Journal of Environmental Pathology, Toxicology and Oncology*, 20(1), 1–7.

Wilcock, A., Pun, M., Khanona, J., Aung, M. 2004. Consumer attitudes, knowledge and behaviour: A review of food safety issues. *Trends in Food Science and Technology*, 15(2), 56–66.

Yu, L., Reitmeier, C.A., Love, M.H. 1996. Strawberry texture and pectin content as affected by electron beam irradiation. *Journal of Food Science*, 61(4), 844–846.

Chapter 9

Bio-Preservation of Dairy Products

A Non-Thermal Processing and Preservation Approach for Shelf-Life Extension of Dairy Products

Nilesh Prakash Nirmal and Chalat Santivarangkna

CONTENTS

9.1 Introduction	111
9.2 Milk and Milk Products and the Chemical Preservatives Used	112
9.3 Bio-Preservation (Non-Thermal Process)	115
9.3.1 Bacteriocins Classifications	115
9.4 Application of Bacteriocins in Dairy Products	116
9.4.1 Nisin	116
9.4.2 Pediocins	119
9.4.3 Lacticins	120
9.4.4 Enterocins	120
9.5 Conclusion	121
References	121

9.1 INTRODUCTION

Milk and milk products (dairy) are highly perishable owing to their high moisture and nutrition content, as well as their neutral pH. In addition, the abundant content of proteins, carbohydrates, lipids, vitamins and minerals make them a suitable vehicle for foodborne pathogens (Delorme et al. 2020). Hence, milk is commonly processed by pasteurization before direct consumption or preparation of milk products (Gálvez et al. 2014). Although thermal processes kill pathogens and reduce some spoilage microorganisms, they also destroy the nutritional and sensorial characteristics of milk products (Singh 2018). Moreover, thermal processes consume energy and cost, which ultimately affect the cost of the final product as well as industry profit (Barba et al. 2017). Therefore, increasing consumer demand for the original taste, flavor and nutritional components of the milk product has forced industry toward non-thermal processes. Additionally, non-thermal processes require less energy and provide food safety, hence securing industry profit (Delorme et al. 2020). Non-thermal processes used to extend the shelf-life of dairy products include high-pressure processing, pulsed electric field, UV light radiation, ultrasound, atmospheric cold plasma, microfiltration and bio-preservation (Barba et al. 2017; Negash and Tsehai 2020; Kareem and Razavi 2020).

Bio-preservation is a natural method to control food pathogens that uses natural microbiota, protective starter cultures or their metabolites to extend the shelf-life of food products (Ameer et al. 2019; Lynch et al. 2021; Singh 2018). Bio-preservatives are more reliable than chemicals for enhancing the quality of milk products by retaining the natural organoleptic properties as well as nutritional compounds (Riesute et al. 2021; Singh 2018). Moreover, consumer demand for chemical-free, minimal process and natural food products make the bio-preservation technique a more promising, hygienic and acceptable process (Madi and Boushaba 2017). Bio-preservatives should not alter the sensorial properties of food or produce a toxic effect. In this context, lactic acid bacteria (LAB) and their metabolites are proven to be more effective, reliable and studied microorganisms (Ameer et al. 2019; Riesute et al. 2021). LAB are generally recognized as safe by the Food and Drug Administration (Negash and Tsehai 2020; Singh 2018).

Hence, the objective of this chapter is to summarize and discuss the recent research findings on the bio-preservation of dairy products. The main focus has been given to bacteriocin, including nisin, pediocin, lacticin and enterocins and their applications in dairy products.

9.2 MILK AND MILK PRODUCTS AND THE CHEMICAL PRESERVATIVES USED

Milk is considered a whole food with numerous nutrients and the single source of nourishment for the newborn (Adeniji and Eyinla 2019). Milk is obtained from a variety of domestic mammals such as cows, buffalo, goats, sheep, camels and humans (Early 2012). Cow milk is used all over the world for human consumption, followed by buffalo, goat, sheep and camel. The major component of cow milk is water (87.3%), milk solids (12.7%), fats (3.7%), protein (3.4%) and minerals (0.7%) (Early 2012). Milk is the raw material for the production of all dairy products including butter, cheese, yogurt, creams, etc. Dairy products or ingredients are commonly used in most food products. Nutritional and physicochemical characteristics of dairy products provide excellent conditions for bacterial growth such as pathogenic and food spoilage bacteria (Gálvez et al. 2014). The growth of spoilage or pathogenic microorganisms always has a negative impact on dairy products by destroying the quality and shelf-life. For instance, psychro-tolerant bacteria can grow during cold storage of milk. In addition, several microorganisms that are known to contaminate milk have an adverse effect on health, causing severe illness.

The most seriously concerning pathogenic microorganisms and the majority of pathogens related to milk or milk-based products are *Campylobacter*, *Escherichia coli* (verocytotoxin producing strain), *Staphylococcus aureus*, *Salmonella* and *Listeria monocytogenes* (Verraes et al. 2015). In foodborne outbreaks worldwide, 2–6% of foodborne infections are related to milk and milk-based products (Claeys et al. 2013). Therefore, milk quality and safety are prime issues for industry and priority research topics. Hence, thermal (traditional) and non-thermal (emerging) processes are used to eliminate the contaminant bacteria from milk and milk products.

A non-thermal process used by the dairy industry is chemical preservatives (Table 9.1), especially for cheeses, and the common chemicals are benzoate, sorbate and natamycin. However, the addition of these preservatives is restricted by the regulation authorities. The use of prohibited chemicals is a big challenge for the milk industry, and it poses concerns regarding the safety and quality of milk and milk products. Examples of harmful preservatives or adulterants commonly used are formaldehyde/formalin, boric acid and antibiotics. Hence bio-preservatives are in demand.

TABLE 9.1 CHEMICAL PRESERVATIVES COMMONLY USED OR BEING OF INTEREST FOR THE MILK INDUSTRY

Chemical Preservatives	Applications and Limitations	Acceptable Daily Intake (ADI)	References
Hydrogen peroxide (H_2O_2)	Hydrogen peroxide is used as a preservative in some countries for milk intended for use in cheese making. The residue of (H_2O_2) can affect cheese quality due to the inhibition of lactic acid bacteria, increase in casein proteolysis with renin and bitter and acid flavors after aging	Hydrogen peroxide (less than 0.05% of the milk weight) is approved in the USA for treating milk used for the production of cheeses. ADI is not allocated	Hanway et al. (2005), JECFA (2004)
Benzoic acid/benzoate	The maximum recommended level of 300 mg/kg of benzoic acid is acceptable for use in flavored fermented milk heat-treated after fermentation and flavored drinks based on fermented milk heat-treated after fermentation	Acceptable daily intake (ADI) of benzoic acid is 0–5 mg/kg bw.	Codex Alimentarius (2011), JECFA (1996)
Sorbic acid/sorbate	The maximum recommended level of sorbates including sorbic acid is 1,000 mg/kg in cheese, flavored fermented milk heat-treated after fermentation and flavored drinks based on fermented milk heat-treated after fermentation; in dairy fat spread, 2,000 mg/kg, singly or in combination for fat contents < 59% and 1,000 mg/kg singly or in combination for fat contents ≥ 59%; and 1,000–3,000 mg/kg depending on cheese types and application (e.g. treatment of cheese surface)	ADI of sorbic acid is 0–25 mg/ kg bw	Codex Alimentarius (2011), JECFA (1996)
Propionic acid/propionate	The maximum level of 3,000 mg/kg of propionic acid is recommended for surface or rind treatment of cheese. The use in cheese mass of unripened cheeses such as cottage, fresh and cream cheeses is limited by GMP	ADI for propionic acid and its sodium, potassium and calcium salts is "not limited" by JECFA. EFSA concluded that potassium propionate could be added to the list of preservatives and established an ADI "not specified"	Codex Alimentarius (2011), JECFA (1974), EFSA (2014)

(Continued)

TABLE 9.1 (CONTINUED) CHEMICAL PRESERVATIVES COMMONLY USED OR BEING OF INTEREST FOR THE MILK INDUSTRY

Chemical Preservatives	Applications and Limitations	Acceptable Daily Intake (ADI)	References
Sodium and potassium nitrate	The maximum recommended level is 35 mg/kg singly or in combination (expressed as nitrate ion) in cheeses	ADI of nitrate is 0–3.7 mg/kg bw (expressed as nitrate ion)	Codex Alimentarius (2011), JECFA (2002a)
Natamycin (pimaricin)	Natamycin may be used for the surface or rind treatment of semi-hard and semi-soft cheeses only. The maximum recommended level is 2 mg/dm² of surface, not present in a depth of 5 mm	ADI of natamycin is 0–3.7 0.3 mg/kg bw. The issue of antimicrobial resistance should be addressed	Codex Alimentarius (2011), JECFA (2002b)
Carbon dioxide	It retards the growth of bacteria in raw milk and reduces the degree of proteolysis and lipolysis of pasteurized milk. However, CO_2 should be removed before the pasteurization to minimize deposits on the walls of the pasteurizer	No clear evidence suggests that CO_2 in food or drinks per se is harmful	Hotchkiss et al. (2006)
Nitrogen gas	The use of N_2 gas in raw milk has shown the potential to control bacterial growth in a cool temperature range without undesired side effects	The complete evaporation of liquid nitrogen normally occurs before reaching the consumer. However, the US FDA warns consumers of the potential for serious injury from eating, drinking or handling food products prepared by adding liquid nitrogen immediately before consumption	Dechemi et al. (2005); Munsch-Alatossava et al. (2016)

9.3 BIO-PRESERVATION (NON-THERMAL PROCESS)

Bio-preservation is a sustainable strategy for the shelf-life extension of dairy products during manufacturing and storage. Bio-preservation of dairy products is conducted using the metabolites produced by inherent microorganisms in milk such as LAB (Ameer et al. 2019). LAB are the most dominant bacteria present in raw milk as they largely depend on lactose for their growth and metabolism (Madi and Boushaba 2017). The most common genera of LAB present in milk include *Enterococcus*, *Lactobacillus*, *Lactococcus*, *Leuconostoc*, *Psychrotrophic* and *Streptococcus*, which predominantly get established during cold storage (Quigley et al. 2013). Further, the growth of these microorganisms can directly affect the development of any dairy product. Bacteriocins that are reported to improve the shelf-life of dairy products include nisin, pediocins, lacticins and enterocins (Silva et al. 2018). Bacteriocins from LAB effectively inhibit a broad spectrum of pathogenic microorganisms and are widely used in the dairy industry because of LAB's GRAS status (Meng et al. 2021; Silva et al. 2018; Ameer et al. 2019).

9.3.1 Bacteriocins Classifications

Bacteriocins are ribosomal synthesized bactericidal peptides or complex proteins with antimicrobial action against diverse groups of undesirable closely related bacteria (Nath et al. 2014). Bacteriocins are different from medicinal antibiotics in that they can be easily digested by proteases in the human digestive tract and hence do not affect or alter the digestive microbiota (Negash and Tsehai 2020). Bacteriocins not only prevent spoilage and pathogenic bacteria but also promote useful bacterial populations (Kareem and Razavi 2020). Moreover, a small concentration or amount of bacteriocin can kill or inhibit spoilage and pathogenic bacteria that compete for the same nutrient medium (Nath et al. 2014; Negash and Tsehai 2020). Therefore, bacteriocins could be a promising antimicrobial candidate for food and feed producers. Additionally, bacteriocins from LAB have many other advantages which make them suitable candidates for dairy processing such as high thermo-stability, wide pH range, inactivity against eukaryotic cells, non-toxicity and no adverse effects on product quality (Riesute et al. 2021; Strack et al. 2020). Bacteriocins produced by LAB are cationic peptides with hydrophobic patches and contain up to 60 amino acid residues (Singh 2018).

Bacteriocins from LAB have been classified into three major classes depending on their characteristics and modes of action (Table 9.2). Class-I is the peptides that undergo enzymatic modification after ribosomal synthesis. These are heat-stable peptides with a molecular weight of less than 10 kDa. Generally, class-I bacteriocins contain different thioether-based intramolecular rings of lanthionine and β-methyl lanthionine (Singh 2018). Class-I is further subdivided into Ia, Ib, Ic, Id and Ie, depending on their mode of action. Class-II bacteriocins are unmodified peptides with less than 10 kDa. They are normally cationic peptides with high isoelectric points and are also heat stable. They have an amphiphilic helical structure, which allows them to interact with the cell membrane of the target microorganism (Negash and Tsehai 2020). Class-II is subdivided into IIa, IIb, IIc and IId based on different N-terminal sequences (Alvarez-Sieiro et al. 2016). Lastly, class-III bacteriocins are larger than 10 kDa and heat-sensitive protein. These are unmodified proteins with bacteriolytic or non-lytic activity (Alvarez-Sieiro et al. 2016; Negash and Tsehai 2020). These bacteriocins directly act on the cell wall of target bacteria, leading to the lysis of the cell (Nath et al. 2014).

TABLE 9.2 CLASSIFICATION OF BACTERIOCINS FROM LAB

Type	Characteristics	Subcategories	Examples
Class-I	These peptides are ribosomally produced and post-translationally modified, heat stable, molecular weight < 10 kDa	Ia – lanthipeptides Ib – cyclized peptides Ic – linear azol(in)e-containing peptide Id – sactibiotic Ie – glycocin If – lasso peptide	Nisin-A Enterocin AS-48 Streptolysin S Subtilosin A Glycocin F Microcin J25
Class-II	Ribosomally produced unmodified peptides, heat stable, molecular weight < 10 kDa	IIa – pediocin-like IIb – two peptides IIc – leaderless IId – non-pediocin-like single peptide	Pediocin PA-1 Lactococcin Q Lacticin Q, Enterocin L-50 Laterosporulin
Class-III	Unmodified, heat-sensitive protein, molecular weight > 10 kDa	IIIa – bacteriolysins IIIb – non-lytic	Zoocin A Helveticin J

Sources: Alvarez-Sieiro et al. (2016), Negash and Tsehai (2020).

9.4 APPLICATION OF BACTERIOCINS IN DAIRY PRODUCTS

9.4.1 Nisin

Nisin is produced by *Lactococcus lactis* subsp. *lactis* and composed of 34 amino acid residues (Sobrino-López and Martín-Belloso 2008). It has a molecular weight of 3.5 kDa and is classified as a Class-Ia (lantibiotic). Nisin is declared safe by both the Food and Agriculture Organization (FAO) and the World Health Organization (WHO) expert committee on food preservatives (Silva et al. 2018). It is commercially available as Nisaplin™ and used in over 48 countries (Negash and Tsehai 2020). It has antimicrobial activity against a wide range of Gram-positive bacteria, including LAB and pathogenic microorganisms such as *Staphylococcus* and *Listeria* (Gálvez et al. 2014). Additionally, it becomes effective in reducing or killing heat-resistant spore-forming bacteria including *Bacillus* and *Clostridium* (Ibarra-Sánchez et al. 2020).

The antimicrobial mechanism of nisin involves binding with a peptidoglycan precursor lipid II of the target cell and forming a pore as well as inhibiting the cell wall formation (Lubelski et al. 2008; Alvarez-Sieiro et al. 2016). Nisin is effective and widely used in cheese and spread processing, where it protects from late blowing caused by spore-producing bacteria like *Clostridium* (Ameer et al. 2019; Negash and Tsehai 2020). Additionally, it reduces and prevents the occurrence of pathogenic bacteria, including *L. monocytogenes* and *S. aureus* in milk and milk products (Madi and Boushaba 2017; Pinto et al. 2011; Arqués et al. 2011). Table 9.3 represents the recent investigation of various bacteriocins alone or in combination with other hurdles to inhibit pathogenic and spoilage microorganisms in dairy products. The effect of nisin at various concentrations (0, 100 and 500 IU/ml) on *S. aureus* surveillance in traditional Minas Serro cheese was studied by Pinto et al. (2011). They found that reduction in *S. aureus* count was dose-dependent on the nisin concentration used, where 100 IU/ml inhibited a 1.2 log cycle, while 500 IU/ml inhibited a 2 log cycle of *S. aureus* on day seven. The reduction in *S. aureus* was correlated with the bactericidal effect of the nisin. Moreover, the addition of nisin improved the physicochemical and mechanical properties of Serro cheese (Pinto et al. 2011). In another study, the application of

9.4 Application of Bacteriocins in Dairy Products 117

TABLE 9.3 APPLICATIONS OF VARIOUS BACTERIOCINS IN DAIRY PRODUCT QUALITY CONTROL

Bacteriocins	Bacterial Strain/ Additional Hurdle	Dairy Product	Applications/Effective Concentration	References
Nisin	NA	Traditional Minas Serro cheese	Reduction of pathogenic *S. aureus* 500 IU/ml (2.0 log reduction)	Pinto et al. (2011)
	Nisaplin	High-fat milk pudding	Control of spoilage bacteria (*B. thuringiensis*, *B. cereus* and *Paenibacillus jamilae*) 80 and 120 IU/g for 5.0 and 7.5% fat milk pudding, resp.	Oshima et al. (2014)
	Lc. Lactis DF4Mi	Minas-type goat cheese	Control of *L. monocytogenes*. Bacteriostatic effect (3 log reduction)	Furtado et al. (2015)
	Nisin plus garlic extract nanoliposomes	Whole milk	Antimicrobial effect against pathogenic Gram-positive and Gram-negative bacteria (*E. coli*, *S. aureus*, *S. enteritidis* and *L. monocytogenes*)	Pinilla and Brandelli (2016)
	Nisin plus *Ziziphora clinopodioides* essential oil (ZEO)	Whole milk	Reduction in *S. aureus* and *S. typhimurium*. ZEO (0.2%) + Nisin (500 IU/ml)	Shahbazi (2016)
	Nisin and high-pressure carbon dioxide	Dairy product	Inactivation of *E. coli* and *S. aureus*	Li et al. (2016)
	Lc. lactis	Cheese	Anti-*Listeria* activity	Ho et al. (2018)
	Lc. lactis	Fresh cheese	Anti-*Listeria* activity	Kondrotiene et al. (2018)
	NA	Whole milk	Inhibitory effect against *Mycobacterium avium* ssp. *paratuberculosis* (500 IU/ml)	Ali et al. (2019)
	Nisin plus lysozyme	Low-salt soft cheese	Reduction of aerobic spore-forming bacteria nisin (25 mg/kg) + lysozyme (100 mg/kg)	Shaala et al. (2020)
Pediocins	Eight antagonistic strains. Cell-free supernatants	Milk	Anti-*Listeria* activity	Hartmann et al. (2011)
	Fermented cheese whey	Buffalo milk	Reduction of total viable counts	Verma et al. (2017)

(*Continued*)

TABLE 9.3 (CONTINUED) APPLICATIONS OF VARIOUS BACTERIOCINS IN DAIRY PRODUCT QUALITY CONTROL

Bacteriocins	Bacterial Strain/ Additional Hurdle	Dairy Product	Applications/Effective Concentration	References
Lacticins	*Lc. Lactis* subsp. *lactis* INIA 415 (lacticin 481)	Ovine milk cheese	Inhibition of *Clostridium beijerinckii* spores and late blowing in cheese	Garde et al. (2011)
	Lc. Lactis 32FL1 and 32FL3 strains (lacticin 481)	Cottage cheese	Reduction of *L. monocytogenes* growth	Dal Bello et al. (2012)
	Lc. Lactis L3A21M1 (lacticin 481)	Azorean Pico cheese	Anti-*Listeria* activity	Ribeiro et al. (2016)
	Lc. Lactis IFPL 3593 (lacticin 3147)	Cheese	Inhibit *Clostridium* spore and late blowing in cheese	Martínez-Cuesta et al. (2010)
Enterocins	*E. mundtii* CRL35	Fresh Minas cheese	Inhibition of pathogenic and non-pathogenic microorganism. Control of *L. monocytogenes* in cheese	Pingitore et al. (2012)
	E. faecium strains	Fresh whey cheese	Control the growth of *L. monocytogenes*	Aspri et al. (2017)
	E. faecium A15	Egyptian dairy products	Anti-*Listeria* activity	El-Ghaish et al. (2017)
	E. faecalis L3A21M3 and L3B1K3	Cheese	Reduction of *L. monocytogenes* in a dose-dependent manner	Ribeiro et al. (2017)

nisin or a bacteriocinogenic *La. lactis* DF04Mi strain in fresh Minas-type goat cheese was investigated for *L. monocytogenes* inactivation (Furtado et al. 2015). Authors reported that the addition of nisin (12.5 mg/kg) rapidly diminished the *L. monocytogenes* count, whereas the DF04Mi strain controlled the *L. monocytogenes* count from day zero until day ten of refrigerated storage. They further suggested that the bacteriocinogenic DF04Mi strain should be studied for nisin isolation and purification (Furtado et al. 2015). Similarly, Kondrotiene et al. (2018) and Ho et al. (2018) isolated numerous bacteriocinogenic *La. Lactis* strains from different milk and vegetable sources, respectively. Kondrotiene et al. (2018) reported that only four strains encoding nisin z (22, 56, 59 and 63) were safe regarding enzymatic activities, antibiotic susceptibility and food processing characteristics. Meanwhile, Ho et al. (2018) suggested that even though 14 wild isolated *Lc. Lactis* strains possessed strong anti-*Listeria* activity, a further study on their virulence factors and biogenic amines formations needed to be conducted.

The antibacterial activity of the nisin can be improved by combining the effect of nisin with other hurdles (Kondrotiene et al. 2018). In this context, the combined effects of nisin with garlic extract (Pinilla and Brandelli 2016), nisin plus *Ziziphora clinopodioides* essential oil (Shahbazi 2016) and nisin in combination with lysozymes (Shaala et al. 2020) showed that the combined effects significantly lowered Gram-positive, Gram-negative and aerobic spore-forming bacteria compared to nisin alone. Additionally, the synergetic effect of nisin with high-pressure carbon dioxide (HPCD) was investigated on the inactivation of *E. coli* and *S. aureus* (Li et al. 2016). Authors reported that nisin in combination with HPCD damaged the outer membrane, then the cytoplasmic membrane, leading to cell lysis. Although nisin has broad antibacterial activity against pathogenic and spoilage bacteria, its efficacy was found to be affected by pH, temperature, food composition and microbiota (Zhou et al. 2014). Hence, the search for other stable and active bacteriocins is still a research priority for scientists.

9.4.2 Pediocins

Pediocins are produced by *Pediococcus* spp. and classified as class-IIa bacteriocins. Pediocin LB-B1 was produced and isolated from *L. plantarum* and showed molecular weight in the range of 2.5–6.5 kDa (Xie et al. 2011). This pediocin LB-B1 was heat stable at 121° C for 15 min with a broad pH range (2.0–10.0) but was sensitive to proteolytic enzymes. Pediocin LB-B1 had a bactericidal effect against *Listeria*, *Lactobacillus*, *Streptococcus*, *Enterococcus*, *Pediococcus* and *Escherichia*. Further gene cluster encoding revealed that LB-B1 had 99.8% similarity with pediocin PA-1 (Xie et al. 2011). In another study, *P. acidilactici* OB4 and *P. pentosaceus* SM3 isolated from *ogi*, fermented cow and sheep milk showed tolerance towards bile salts and bile salt hydrolytic activity (Banwo et al. 2013). Further, pediocins produced by these two isolates were stable at pH 4–9 and treatment with lipase, catalase, α-amylase and lysozyme. Moreover, pediocin (PedA) produced by OB4 was stable at 100° C for 30 min, whereas pediocin (PedB) produced by SM3 was stable for 10 min. Both of the pediocins had high anti-*Listeria* activity and antibacterial activity against some pathogenic microorganisms (Banwo et al. 2013). Although pediocins showed good heat stability, proteolytic stability and a wide pH range, very few studies have been conducted for the application of pediocins in dairy products.

Cell-free supernatants containing pediocin PA-1 showed higher anti-*Listeria* activity in milk compared to other bacteriocins (Hartmann et al. 2011). However, this activity was only for a short time, and surviving bacteria grew after prolonged storage. This indicates that another hurdle in combination with pediocin is needed to improve the antibacterial activity and shelf-life extension of the milk product (Hartmann et al. 2011). Pediocin PA-1 containing fermented cheese whey showed the highest activity against *S. aureus* and LAB, while being less effective against coliform, yeast and molds in raw buffalo milk (Verma et al. 2017).

9.4.3 Lacticins

Lacticins are class-I bacteriocins and come under the lantibiotic (Ia) group, where nisin is considered as lantibiotic I and lacticin as lantibiotic II (Alvarez-Sieiro et al. 2016). Lacticins are produced by *Lc. lactis* strains and comprised of lacticin 481 (Dal Bello et al. 2012) and lacticin 3147 (Draper et al. 2013). Lantibiotic II are most effective against pathogenic and spore-forming bacteria (Ribeiro et al. 2016; Garde et al. 2011).

Lacticin 481 produced by *Lc. lactis* subsp. *lactis* INIA 415 showed inhibition of *Clostridium beijerinckii* spores in bovine milk cheese (Garde et al. 2011). Moreover, this strain produced lactic acid and other volatile compounds, which are an additional perk for the quality attribute of cheese. Hence, the addition of the *Lc. lactis* subsp. *lactis* INIA 415 strain in a starter culture could prevent the late blowing effect by inhibiting *C. beijerinckii* spore formation in cheese without affecting its organoleptic properties (Garde et al. 2011). In their study, Ribeiro et al. (2016) isolated anti-listerial strain *Lc. Lactis* L3A21M1 from Azorean Pico cheese. The isolated strain produced lacticin 481 with a molecular mass of 2.9 kDa. The produced lacticin was thermostable, active against a wide pH range with broad antimicrobial activity. In addition, lacticin 481 had a bacteriostatic effect against *L. monocytogenes* both in culture media and in a cheese model (Ribeiro et al. 2016). Similarly, Dal Bello et al. (2012) showed that lacticin 481-producing *Lc. Lactis* strains 32Fl1 and 32FL3 effectively controlled the growth of pathogenic bacteria, particularly *L. monocytogenes* in cottage cheese.

Lacticin 3147 is a two-peptide lantibiotic (LtnA1 and LtnA2) with antibacterial activity against various Gram-positive bacteria (Bakhtiary et al. 2017). The antibacterial activity of lacticin 3147 could be enhanced to inhibit Gram-negative bacteria by synergistic application with polymyxin (Draper et al. 2013). Bakhtiary et al. (2017) studied the antibacterial mode of action of the two-peptide lacticin 3147 and found that LtnA1 binds to the peptidoglycan precursor lipid II via C-terminal and inhibits peptidoglycan synthesis, whereas the combination of LtnA1 and LtnA2 can induce rapid membrane lysis of the cell without binding to the Lipid II precursor (Bakhtiary et al. 2017). The lacticin 3147-producing *Lc. lactis* IFPL 3593 strain showed inhibitory activity against germination of *Clostridia* spore, thereby preventing late blowing in semi-hard cheese (Martínez-Cuesta et al. 2010).

9.4.4 Enterocins

Enterocins can be classified as modified (Class-Ib cyclized peptide) and unmodified (Class-IIc leaderless) bacteriocin (Alvarez-Sieiro et al. 2016). *Enterococcus* strains (*E. mundtii* CRL35 and *E. faecium* ST88ch) isolated from cheese showed antimicrobial activity against various pathogenic and non-pathogenic microorganisms (Pingitore et al. 2012). However, the activity was affected by various process conditions such as pH, temperature and the presence of salts and surfactants. Both strains showed a bacteriostatic effect against *L. monocytogenes*. Nevertheless, strain CRL35 inhibited *L. monocytogenes* 426 in cheese for up to 12 days compared to strain ST88ch for six days at 8° C (Pingitore et al. 2012). In another study, enterocin-producing *E. faecalis* strains (L3A21M3 and L3B1K3) were isolated from artisanal cheese and tested against *L. monocytogenes* (Ribeiro et al. 2017). Enterocins produced by these strains reduced the growth of *L. monocytogenes* on fresh cheese with the bacteriostatic effect in a dose-dependent manner. Additionally, both of these strains tolerated gastrointestinal conditions and adhered to human intestinal cells, while blocking the adhesion and invasion of *L. monocytogenes*. However, enterocins produced by both strains were found to be heat- and extreme-pH-sensitive (Ribeiro et al. 2017).

In their study, Aspri et al. (2017) isolated three potential enterocin-producing *E. faecium* strains from donkey milk. All strains were effective against *L. monocytogenes*, *S. aureus* and *B. cereus*. One strain (DM 33) showed a bactericidal effect while the other two (DM 224, 270) showed bacteriostatic effects. Mass spectrometry analyses revealed that peptides produced by these strains were matched with enterocins A and B. Nevertheless, peptides were active over a wide pH range and heat treatment. Additionally, these enterocins were effective in controlling the post-process growth of *L. monocytogenes* in fresh whey cheese (Aspri et al. 2017). *E. faecium* A15 isolated from some traditional Egyptian dairy products showed anti-*Listera* activity against *L. monocytogenes* EGDEe 10776 (El-Ghaish et al. 2017). Polymerase chain reaction analyses revealed the presence of an enterocin B gene in the A15 strain. Enterocin produced from this strain was stable at pH 5–8 and 100° C for 10 min. Hence, isolated *E. faecium* A15 could be used as a bio-preservative to increase the shelf-life of dairy products (El-Ghaish et al. 2017).

9.5 CONCLUSION

Bio-preservation as a non-thermal approach offers a natural and safe way to extend the storage and safety of food products using microbiota and/or their antimicrobial peptides such as bacteriocins. Lactic acid bacteria and their bacteriocins are the most studied and commercialized bio-preservatives for the quality and shelf-life extension of dairy products. Well-known and commercialized bacteriocins include nisins, pediocins, lacticins and enterocins. All of these bacteriocins have antimicrobial activity against pathogenic, spore-forming and food spoilage bacteria, particularly *L. monocytogenes*, *S. aureus*, *B. cereus* and *Clostridium sp*. Most of the bacteriocins show the bacteriostatic effect, while few of them are bactericidal. Nevertheless, their activity is affected by process conditions such as pH, temperature and other food ingredients. However, none of the bacteriocins alone is able to completely eliminate pathogens from the food system. Therefore, the addition of another hurdle or combination of other antimicrobial agents is recommended for the long shelf-life and safety of dairy products.

REFERENCES

Abdulhussain Kareem, Raghda Abdulhussain, and Seyed Hadi Razavi. 2020. "Plantaricin bacteriocins: As safe alternative antimicrobial peptides in food preservation—A review." *Journal of Food Safety* 40(1):e12735. doi: 10.1111/jfs.12735.

Adeniji, P. O., and T. E. Eyinla. 2019. "Microbial growth and changes in nutritional contents of preserved and unpreserved fresh milk stored at 4°C." *Journal of Food and Nutrition Research* 7(6):443–446. doi: 10.12691/jfnr-7-6-5.

Ali, Zeinab I., Adel M. Saudi, Ralph Albrecht, and Adel M. Talaat. 2019. "The inhibitory effect of nisin on *Mycobacterium avium* ssp. *paratuberculosis* and its effect on mycobacterial cell wall." *Journal of Dairy Science* 102(6):4935–4944. doi: 10.3168/jds.2018-16106.

Alvarez-Sieiro, P., M. Montalbán-López, D. Mu, and O. P. Kuipers. 2016. "Bacteriocins of lactic acid bacteria: Extending the family." *Applied Microbiology and Biotechnology* 100(7):2939–2951. doi: 10.1007/s00253-016-7343-9.

Ameer, S., S. Aslam, and M. Saeed. 2019. "Preservation of milk and dairy products by using biopreservatives." *Middle East Journal of Applied Science & Technology* 2(4):72–79.

Arqués, Juan L., Eva Rodríguez, Manuel Nuñez, and Margarita Medina. 2011. "Combined effect of reuterin and lactic acid bacteria bacteriocins on the inactivation of food-borne pathogens in milk." *Food Control* 22(3–4):457–461. doi: 10.1016/j.foodcont.2010.09.027.

Aspri, Maria, Paula M. O'Connor, Des Field, Paul D. Cotter, Paul Ross, Colin Hill, and Photis Papademas. 2017. "Application of bacteriocin-producing *Enterococcus faecium* isolated from donkey milk, in the bio-control of *Listeria monocytogenes* in fresh whey cheese." *International Dairy Journal* 73:1–9. doi: 10.1016/j.idairyj.2017.04.008.

Bakhtiary, Alireza, Stephen A. Cochrane, Pascal Mercier, Ryan T. McKay, Mark Miskolzie, Clarissa S. Sit, and John C. Vederas. 2017. "Insights into the mechanism of action of the two-peptide lantibiotic lacticin 3147." *Journal of the American Chemical Society* 139(49):17803–17810. doi: 10.1021/jacs.7b04728.

Banwo, K., A. Sanni, and H. Tan. 2013. "Functional properties of *Pediococcus* species isolated from traditional fermented cereal gruel and milk in Nigeria." *Food Biotechnology* 27(1):14–38. doi: 10.1080/08905436.2012.755626.

Barba, Francisco J., Mohamed Koubaa, Leonardo do Prado-Silva, Vibeke Orlien, and Anderson Sant'Ana. 2017. "Mild processing applied to the inactivation of the main foodborne bacterial pathogens: A review." *Trends in Food Science and Technology* 66:20–35. doi: 10.1016/j.tifs.2017.05.011.

Carmen Martínez-Cuesta, M. C., José Bengoechea, Irene Bustos, Beatriz Rodríguez, Teresa Requena, and Carmen Peláez. 2010. "Control of late blowing in cheese by adding lacticin 3147-producing *Lactococcus lactis* IFPL 3593 to the starter." *International Dairy Journal* 20(1):18–24. doi: 10.1016/j.idairyj.2009.07.005.

Claeys, Wendie L., Sabine Cardoen, Georges Daube, Jan De Block, Koen Dewettinck, Katelijne Dierick, Lieven De Zutter, André Huyghebaert, Hein Imberechts, Pierre Thiange, Yvan Vandenplas, and Lieve Herman. 2013. "Raw or heated cow milk consumption: Review of risks and benefits." *Food Control* 31(1):251–262. doi: 10.1016/j.foodcont.2012.09.035.

Codex Alimentarius. 2011. *Milk and Milk Products*, 2nd edition, Codex Alimentarius Commission Joint FAO/WHO Food Standards Programme, Rome, Italy.

Dal Bello, Barbara, Luca Cocolin, Giuseppe Zeppa, Des Field, Paul D. Cotter, and Colin Hill. 2012. "Technological characterization of bacteriocin producing *Lactococcus lactis* strains employed to control *Listeria monocytogenes* in cottage cheese." *International Journal of Food Microbiology* 153(1–2):58–65. doi: 10.1016/j.ijfoodmicro.2011.10.016.

Dechemi, S., H. Benjelloun, and J. -M. Lebeault. 2005. "Effect of modified atmospheres on the growth and extracellular enzymes of psychrotrophs in raw milk." *Engineering in Life Sciences* 5(4):350–356.

Delorme, Mariana M., Jonas T. Guimarães, Nathália M. Coutinho, Celso F. Balthazar, Ramon S. Rocha, Ramon Silva, Larissa P. Margalho, Tatiana C. Pimentel, Marcia C. Silva, Monica Q. Freitas, Daniel Granato, Anderson S. Sant'Ana, Maria Carmela K. H. Duart, and Adriano G. Cruz. 2020. "Ultraviolet radiation: An interesting technology to preserve quality and safety of milk and dairy foods." *Trends in Food Science and Technology* 102:146–154. doi: 10.1016/j.tifs.2020.06.001.

Draper, Lorraine A., Paul D. Cotter, Colin Hill, and R. Paul Ross. 2013. "The two peptide lantibiotic lacticin 3147 acts synergistically with polymyxin to inhibit Gram negative bacteria." *BMC Microbiology* 13(1):212. doi: 10.1186/1471-2180-13-212.

Early, R. 2012. "Dairy products and milk-based food ingredients." In: *Natural Food Additives, Ingredients and Flavourings*, edited by D. Baines, and R. Seal, 417–445), Cambridge, England.

EFSA Panel on Food additives and Nutrient Sources added to Food (ANS). 2014, "Scientific Opinion on the re-evaluation of propionic acid (E 280), sodium propionate (E 281), calcium propionate (E 282) and potassium propionate (E 283) as food additives." *EFSA Journal* 12(7):1831.

El-Ghaish, S., M. Khalifa, and A. Elmahdy. 2017. "Antimicrobial impact for *Lactococcus lactis* subsp. *lactis* A15 and *Enterococcus faecium* A15 isolated from some traditional Egyptian dairy products on some pathogenic bacteria." *Journal of Food Biochemistry* 41(1):e12279. doi: 10.1111/jfbc.12279.

Furtado, Danielle N., Svetoslav D. Todorov, Mariza Landgraf, Maria T. Destro, and Bernadette D. G. M. Franco. 2015. "Bacteriocinogenic *Lactococcus lactis* subsp. *lactis* DF04Mi isolated from goat milk: Application in the control of *Listeria monocytogenes* in fresh Minas-type goat cheese." *Brazilian Journal of Microbiology* 46(1):201–206.

Gálvez, Antonio, Rosario Lucas López, Rubén Pérez Pulido, and María José Grande Burgos. 2014. "Biopreservation of milk and dairy products." In: *Food Biopreservation*, edited by Antonio Galvez, María José Grande Burgos, Rosario Lucas López and Rubén Pérez Pulido, 49–69, Springer, New York, NY.

Garde, S., M. Avila, R. Arias, P. Gaya, and M. Nuñez. 2011. "Outgrowth inhibition of *Clostridium beijerinckii* spores by a bacteriocin-producing lactic culture in ovine milk cheese." *International Journal of Food Microbiology* 150(1):59–65. doi: 10.1016/j.ijfoodmicro.2011.07.018.

Hanway, W. H., A. P. Hansen, K. L. Anderson, R. L. Lyman, and J. E. Rushing. 2005. "Inactivation of penicillin G in milk using hydrogen peroxide." *Journal of Dairy Science* 88(2):466–469.

Hartmann, H. Andreas, Thomas Wilke, and Ralf Erdmann. 2011. "Efficacy of bacteriocin-containing cell-free culture supernatants from lactic acid bacteria to control *Listeria monocytogenes* in food." *International Journal of Food Microbiology* 146(2):192–199. doi: 10.1016/j.ijfoodmicro.2011.02.031.

Ho, V. T. T., R. Lo, N. Bansal, and M. S. Turner. 2018. "Characterisation of *Lactococcus lactis* isolates from herbs, fruits and vegetables for use as biopreservatives against *Listeria monocytogenes* in cheese." *Food Control* 85:472–483. doi: 10.1016/j.foodcont.2017.09.036.

Hotchkiss, Joseph H., Brenda G. Werner, and Edmund Y.C. Lee. 2006. "Addition of carbon dioxide to dairy products to improve quality: A comprehensive review." *Comprehensive Reviews in Food Science and Food Safety* 5(4):1541–4337.

Ibarra-Sánchez, Luis A., Nancy El-Haddad, Darine Mahmoud, Michael J. Miller, and Layal Karam. 2020. "Invited review: Advances in nisin use for preservation of dairy products." *Journal of Dairy Science* 103(3):2041–2052. doi: 10.3168/jds.2019-17498.

JECFA (Joint FAO/WHO Expert Committee on Food Additives). 1974. "Toxicological evaluation of some food additives including anticaking agents, antimicrobials, antioxidants, emulsifiers and thickening agents. Propionic acid and its calcium, potassium and sodium salts." Seventeenth Report of the JECFA, WHO Technical Report Series No. 539, Geneva, Switzerland.

JECFA. 1996. "Evaluation of certain food additives and contaminants. Forty-sixth report of the Joint FAO/WHO expert committee on food additives." World Health Organization Technical Report Series, Geneva, Switzerland 868.

JECFA. 2002a. "Evaluation of certain food additives and contaminants." Fifty-Ninth Report of JECFA. Joint FAO/WHO Expert Committee on Food Additives, WHO TechnicalReports Series 913, Geneva, Switzerland.

JECFA. 2002b. "Natamycin. Evaluation of certain food additives." Report of the Joint FAO/WHO Expert Committee on Food Additives, (57th meeting). WHO Food Additives Series 48:49–76.

JECFA. 2004. "Evaluatio of certain food additives." Sixty-third report of the Joint FAO/WHO Expert Committee on Food Additives 204. World Health Organization Technical Report Series 928, World Health Organisation, Geneva, Switzerland.

Kondrotiene, Kristina, Neringa Kasnauskyte, Loreta Serniene, Greta Gölz, Thomas Alter, Vilma Kaskoniene, Audrius Sigitas Maruska, and Mindaugas Malakauskas. 2018. "Characterization and application of newly isolated nisin producing *Lactococcus lactis* strains for control of *Listeria monocytogenes* growth in fresh cheese." *LWT-Food Science and Technology* 87:507–514. doi: 10.1016/j.lwt.2017.09.021.

Li, Hui, Zhenzhen Xu, Feng Zhao, Yongtao Wang, and Xiaojun Liao. 2016. "Synergetic effects of high-pressure carbon dioxide and nisin on the inactivation of *Escherichia coli* and *Staphylococcus aureus*." *Innovative Food Science and Emerging Technologies* 33:180–186. doi: 10.1016/j.ifset.2015.11.013.

Lubelski, J., R. Rink, R. Khusainov, G. N. Moll, and O. P. Kuipers. 2008. "Biosynthesis, immunity, regulation, mode of action and engineering of the model lantibiotic nisin." *Cellular and Molecular Life Sciences* 65(3):455–476. doi: 10.1007/s00018-007-7171-2.

Lynch, David, Colin Hill, Des Field, and Máire Begley. 2021. "Inhibition of *Listeria monocytogenes* by the *Staphylococcus capitis*—Derived bacteriocin capidermicin." *Food Microbiology* 94:103661. doi: 10.1016/j.fm.2020.103661.

Madi, Nassim, and Rihab Boushaba. 2017. "Identification of potential biopreservative lactic acid bacteria strains isolated from Algerian cow's milk and demonstration of antagonism against *S. aureus* in cheese." *Food Science and Technology Research* 23(5):679–688. doi: 10.3136/fstr.23.679.

Meng, Fanqiang, Xiaoyu Zhu, Haizhen Zhao, Ting Nie, Fengxia Lu, Zhaoxin Lu, and Yingjian Lu. 2021. "A class III bacteriocin with broad-spectrum antibacterial activity from *Lactobacillus acidophilus* NX2-6 and its preservation in milk and cheese." *Food Control* 121. doi: 10.1016/j.foodcont.2020.107597. 107597.

Munsch-Alatossava, P., R. Quintyn, I. De Man, T. Alatossava, and J. P. Gauchi. 2016. "Efficiency of N_2 gas flushing compared to the lactoperoxidase system at controlling bacterial growth in bovine raw milk stored at mild temperatures." *Frontiers in Microbiology* 7:839.

Nath, S., S. Chowdhury, K. C. Dora, and S. Sarkar. 2014. "Role of biopreservation in improving food safety and storage." *International Journal of Engineering Research and Application* 4(1):26–32.

Negash, Abebe Worku, and Berhanu Andualem Tsehai. 2020. "Current applications of bacteriocin." *International Journal of Microbiology* 2020:4374891. doi: 10.1155/2020/4374891.

Oshima, S., A. Hirano, H. Kamikado, J. Nishimura, Y. Kawai, and T. Saito. 2014. "Nisin A extends the shelf life of high-fat chilled dairy dessert, a milk-based pudding." *Journal of Applied Microbiology* 116(5):1218–1228. doi: 10.1111/jam.12454.

Pinilla, Cristian Mauricio Barreto, and Adriano Brandelli. 2016. "Antimicrobial activity of nanoliposomes co-encapsulating nisin and garlic extract against Gram-positive and Gram-negative bacteria in milk." *Innovative Food Science and Emerging Technologies* 36:287–293. doi: 10.1016/j.ifset.2016.07.017.

Pinto, M. S., A. F. de Carvalho, A. CdS. Pires, A. A. Campos Souza, P. H. Fonseca da Silva, D. Sobral, J. C. J. de Paula, and A. de Lima Santos. 2011. "The effects of nisin on *Staphylococcus aureus* count and the physicochemical properties of Traditional Minas Serro cheese." *International Dairy Journal* 21(2):90–96. doi: 10.1016/j.idairyj.2010.08.001.

Quigley, Lisa, Orla O'Sullivan, Catherine Stanton, Tom P. Beresford, R. Paul Ross, Gerald F. Fitzgerald, and Paul D. Cotter. 2013. "The complex microbiota of raw milk." *FEMS Microbiology Reviews* 37(5):664–698. doi: 10.1111/1574-6976.12030.

Ribeiro, Susana C., Paula M. O'Connor, R. Paul Ross, Catherine Stanton, and Célia C. G. Silva. 2016. "An anti-listerial *Lactococcus lactis* strain isolated from Azorean Pico cheese produces lacticin 481." *International Dairy Journal* 63:18–28. doi: 10.1016/j.idairyj.2016.07.017.

Ribeiro, Susana C., R. Paul Ross, Catherine Stanton, and Célia C. G. Silva. 2017. "Characterization and application of antilisterial enterocins on model fresh cheese." *Journal of Food Protection* 80(8):1303–1316. doi: 10.4315/0362-028x.Jfp-17-031.

Riesute, Reda, Joana Salomskiene, David Saez Moreno, and Sonata Gustiene. 2021. "Effect of yeasts on food quality and safety and possibilities of their inhibition." *Trends in Food Science and Technology* 108:1–10. doi: 10.1016/j.tifs.2020.11.022.

Shaala, E., S. Awad, and A. M. Nazem. 2020. "Application of natural antimicrobial additives and protective culture to control aerobic spore forming bacteria in low salt soft cheese." *World's Veterinary Journal* 10(4):609–616. doi: 10.29252/scil.2020.wvj73.

Shahbazi, Yasser. 2016. "Effects of ziziphora clinopodioides essential oil and nisin on the microbiological properties of milk." *Pharmaceutical Sciences* 22(4):272–278. doi: 10.15171/ps.2016.42.

Silva, Célia C. G., Sofia P. M. Silva, and Susana C. Ribeiro. 2018. "Application of bacteriocins and protective cultures in dairy food preservation." *Frontiers in Microbiology* 9:594–594. doi: 10.3389/fmicb.2018.00594.

Singh, V. P. 2018. "Recent approaches in food bio-preservation—A review." *Open Veterinary Journal* 8(1):104–111. doi: 10.4314/ovj.v8i1.16.

Sobrino-López, A., and O. Martín-Belloso. 2008. "Use of nisin and other bacteriocins for preservation of dairy products." *International Dairy Journal* 18(4):329–343. doi: 10.1016/j.idairyj.2007.11.009.

Strack, L., R. C. Carli, R. Vd Silva, K. B. Sartor, L. M. Colla, and C. O. Reinehr. 2020. "Food biopreservation using antimicrobials produced by lactic acid bacteria." *Research, Society and Development* 9(8). doi: 10.33448/rsd-v9i8.6666.

Vera Pingitore, E., S. D. Todorov, F. Sesma, and B. D. Franco. 2012. "Application of bacteriocinogenic *Enterococcus mundtii* CRL35 and *Enterococcus faecium* ST88Ch in the control of *Listeria monocytogenes* in fresh Minas cheese." *Food Microbiology* 32(1):38–47. doi: 10.1016/j.fm.2012.04.005.

Verma, S. K., S. K. Sood, R. K. Saini, and N. Saini. 2017. "Pediocin PA-1 containing fermented cheese whey reduces total viable count of raw buffalo (*Bubalis Bubalus*) milk." *LWT—Food Science and Technology* 83:193–200. doi: 10.1016/j.lwt.2017.02.031.

Verraes, C., G. Vlaemynck, S. Van Weyenberg, L. De Zutter, G. Daube, M. Sindic, M. Uyttendaele, and L. Herman. 2015. "A review of the microbiological hazards of dairy products made from raw milk." *International Dairy Journal* 50:32–44. doi: 10.1016/j.idairyj.2015.05.011.

Xie, Ying, Haoran An, Yanling Hao, Qianqian Qin, Y. Huang, Y. Luo, and L. Zhang. 2011. "Characterization of an anti-Listeria bacteriocin produced by Lactobacillus plantarum LB-B1 isolated from koumiss, a traditionally fermented dairy product from China." *Food Control* 22(7):1027–1031. doi: 10.1016/j.foodcont.2010.12.007.

Zhou, Hui, Jun Fang, Yun Tian, and Xiang Yang Lu. 2014. "Mechanisms of nisin resistance in Gram-positive bacteria." *Annals of Microbiology* 64(2):413–420. doi: 10.1007/s13213-013-0679-9.

Chapter 10

Treatment of Dairy Industry Wastewater with Non-Thermal Technologies

Maryam Enteshari and Sergio I. Martinez-Monteagudo

CONTENTS

10.1 Introduction	127
10.2 Sources and Characteristics of Wastewater	128
10.3 Treatments of Wastewater	129
10.3.1 Primary Treatment	130
10.3.2 Secondary Treatment	130
10.4 Non-Thermal Approaches	131
10.4.1 Membrane Processing	131
10.4.2 Hydrodynamic Cavitation	132
10.4.3 Electrochemical Methods	132
10.4.3.1 Electroflotation	132
10.4.3.2 Electrocoagulation	135
10.4.4 Advanced Oxidation Methods	136
10.4.5 Subcritical Water	137
10.5 Conclusions	138
References	139

10.1 INTRODUCTION

Wastewater (WW) is ubiquitous within the dairy industry. The generation of WW starts in the farm after milking the cows, continues in the plant during the manufacture of dairy products, and ends with the products being shipped for distribution and delivery. The dairy industry ranks second in the generation of WW amongst the food sectors (Wang and Serventi 2019). Normal operation of equipment, cleaning protocols, product changeover, and accidental spills contribute to a large amount of WW generated by the dairy industry.

The WW derived from the dairy industry is quite variable in composition and quality since it depends on the specific operation and throughput of the plant. Characteristics and composition of WW generated from fluid milk, butter, cheese, and ice cream have been reviewed elsewhere (Catenacci et al. 2020). Overall, dairy WW contains proteins, fat, carbohydrates, minerals, flavoring compounds, and residual detergents. Such composition significantly contributes to the biological oxygen demand (BOD) and chemical oxygen demand (COD) of WW that is roughly 20 times higher than that of domestic WW (Enteshari and Martínez-Monteagudo 2020).

A confluence of social and economic factors has made the dairy industry consider waste streams as essential actors in the transition to sustainable manufacturing (Enteshari and Martínez-Monteagudo 2018). The reduction and subsequent treatment of WW have become a central issue for the dairy industry. Overall, the treatment of WW requires a combination of several operations, including grease traps, oil–water separators, sedimentation, and biological and chemical treatments. The different treatments used for dairy WW have been reviewed elsewhere (Ahmad et al. 2019). Aerobic and anaerobic treatments are aimed at reducing organic load through the degradation process.

Researchers worldwide are continuously investigating non-thermal technologies that reduce the concentration of pollutants and recover some valuable compounds within the WW stream (peptides, minerals, and fat). Membrane filtration, ultrasound, hydrodynamic cavitation, subcritical water, and advanced oxidation methods are examples of non-thermal technologies.

10.2 SOURCES AND CHARACTERISTICS OF WASTEWATER

Dairy WW within the context of this chapter refers to the residual water after being used in manufacturing operations. Under this definition, any industrial activity that utilizes water contributes toward the generation of WW, including water used for processing, utilities, and cleaning. Table 10.1 outlines the main sources of WW, although the amount and type of pollutants in WW strongly depend on the product being manufactured, which may contain milk product lost, dairy and non-dairy ingredients, starter cultures, byproducts, contaminants, reagents, and cleaning agents.

Most of WW is generated through the different processing steps, including cleaning and washing the floors, vehicles, and trollies, and cleaning-in-place (CIP) protocols. Overall, CIP includes three phases, including 1) the pre-rinse phase, where residuals of raw materials or final products are removed with hot water, 2) the hot-caustic wash, where the solid deposits are removed by

TABLE 10.1 SOURCES OF WASTEWATER WITHIN THE DAIRY INDUSTRY

Source	Comment	Pollutant
Farm	Washing and cleaning of milk parlors	Soil and milk constituents
Transportation and handling	Washing and cleaning of trucks, cans, piping, and tanks	Milk constituents and residual detergents
Manufacture	Spillage produced leak, overflow, equipment malfunction	Non-dairy ingredients and milk constituents
Processing losses	Sludges discharges, product wastes, start-up, shut-down, product changeover	Non-dairy ingredients and milk constituents
Byproduct	Returned products and generation of byproducts	Residual byproduct and cleaning agent
Clean-in-place	Routine cleaning of processing lines	Milk constituents and residual detergents
Equipment performance	Lubricants from machinery	Oils and grease

TABLE 10.2 CHARACTERISTICS OF WASTEWATER FROM DIFFERENT DAIRY PRODUCTS

Dairy Product	COD (g L⁻¹)	BOD (g L⁻¹)	TSS (g L⁻¹)	TS (g L⁻¹)	pH	Reference
Butter	8	0.2–2.6	0.7–5.0	N.R.	12	Janczukowicz, et al. (2008)
Cheese	1–63	0.3–5.0	0.5–3.0	1.9–53	3–9	Gavala et al. (1999)
Ice cream	7–30	8–38	0.3–1.0	15–30	3–8	Enteshari and Martínez-Monteagudo (2020)
Milk	5–10	3.0–5.0	0.1–0.4	3.0–7.0	4–7	Karadag et al. (2015)

COD = chemical oxygen demand; BOD = biological oxygen demand; TSS = total suspended solids; TS = total solids; N.R. = not reported

the action of detergents, and 3) the water rinse phase, where residue of the caustic solution is removed.

Streams of WW are generated intermittently, where the flow rate and content vary significantly with the equipment type and processing cycle. The different pollutants within the WW negatively impact the quality parameters, including BOD, COD, total solids, total suspended solids, total dissolved solids, total volatiles, minerals, and pH. Table 10.2 summarizes the characteristics of WW from different dairy products.

The properties of dairy effluents, such as quantity, physiochemical composition, and biological load, are influenced by the type of product, manufacturing program and method, processing steps, layout of processing facilities in the plant, water management level, and volume of water usage. Thus, there are three main sources of dairy WW:

1) Processing water – the water used for heating and cooling during normal operation. Typically, this effluent type is mostly free of pollutants and chemical agents, requiring the least reuse treatment as runoff water.
2) Cleaning wastewaters – this type of WW is generated through the cleaning protocols in contact with milk and dairy products, leakage of milk products, and water derived from processing equipment failures. This WW contains residuals from milk, cheese, whey, cream, and water from the separator, dilution water from yogurt manufacturing, and fruit particles.
3) Sanitary wastewater – dairy effluents emanate via a cleaning system that encompasses different types of sanitizing agents and detergents. Thus, the pH level of the dairy stream varies noticeably depending on the CIP and cleaning process. Caustic soda, nitric acid, phosphoric acid, and sodium hypochlorite are among the most utilized for the CIP process, influencing the pH of wastewater (Danalewich et al. 1998). The CIP chemicals contribute notably to the levels of biochemical and chemical oxygen demands (BOD and COD) which considerably impact the toxicity level of the waste stream.

10.3 TREATMENTS OF WASTEWATER

Streams of WW are harmful to the environment, and they need to be treated before either direct or indirect discharge. Direct discharge requires removing fats, oils, and grease followed by a biological treatment to reduce the BOD and COD to a permissible level. Indirect discharge involves the treatment of WW at a nearby sewage plant. In general, the treatment of WW requires a combination of several operations, including grease traps, oil–water separators, sedimentation, and biological and chemical treatments. Thus, various arrangements can exist depending on the required reduction of a given pollutant. Despite such variations, the overall process consists of a primary (mechanical) treatment and a secondary (biological and/or chemical) treatment.

10.3.1 Primary Treatment

The separation step is commonly performed with screens, grid traps, and sand traps. Fats, oils, and grease (FOG) are also removed to prevent further blockage of the sewer system (Enteshari and Martínez-Monteagudo 2020). Moreover, the presence of FOG negatively impacts subsequent treatments. Dairy WW exhibits large variations in the pH, ranging from 3 to 10. This is because of the different cleaning protocols used during clean-in-place. Adjustment in the pH is a common pretreatment step before subsequent treatments.

10.3.2 Secondary Treatment

The secondary treatment can be divided into physicochemical and biological methods. Coagulation, flocculation, and adsorption techniques are examples of physicochemical methods, and they have been used for the removal of suspended, colloidal, and dissolved compounds. Coagulation and flocculation are common processes for purification purposes in industrial WW streams. Utilizing chemical coagulants improves the removal of insoluble particles and dissolved organic materials following sedimentation, floatation, and filtration steps (Kushwaha et al. 2011). Numerous coagulation agents such as $FeClSO_4$, H_2SO_4, and carboxymethylcellulose (CMC) have been used to treat dairy streams. Using lactic acid by addition of CMC has shown a promising trend in the chemical treatment of dairy wastewater. However, $FeClSO_4$ has removed 2–3% more COD than H_2SO_4 plus CMS and 4–6% more COD than lactic acid plus CMC (Rusten et al. 1993).

On the other hand, biological treatment is carried out to decompose the organic matter of the WW (Carvalho et al. 2013). The most common biological treatments are trickling filters, aerated lagoons, activated sludge, up-flow anaerobic sludge blankets (UASBs), anaerobic filters, and sequential batch reactors (SBRs). According to oxygen intake, biological treatment can be categorized into two different aerobic and anaerobic techniques. For the aerobic process, microbial degradation and oxidation of waste substances require oxygen, which could be done by activated sludge, trickling filters, aerated lagoons, and/or integration (Kushwaha et al. 2011). Except for proteins and fats, all other substances in WW are biodegradable and can be broken down by the combination of enzymatic and anaerobic treatments.

The sequential batch reactor (SBR) has been recognized as one of the most efficient aerobic methods for treating dairy wastewater. About 90% reduction of COD was reported in wastewater from an industrial milk factory treated with a bench-scale aerobic SBR (Khalaf et al. 2021). The simultaneous application of an aerobic granular activated sludge SBR (GAS-SBR) has shown high efficiency of aerobic treatment with improved settling of residuals for dairy treatment. Anaerobic granular activated sludge SBR has been applied to treat industrial dairy wastewater, resulting in a 90% reduction of total COD, 80% of total nitrogen, and 67% of total phosphorus (Schwarzenbeck et al. 2005).

During the past few decades, anaerobic processes have found promising applications in treating dairy waste streams (Omil et al. 2003). Several drawbacks of aerobic treatment, such as high energy consumption, have motivated researchers to search for other cost-effective processes. Certain characteristics of dairy effluent like high organic content, high COD value, and high temperature have proposed it as a superb matrix for anaerobic treatment. An up-flow anaerobic sludge blanket (UASB) is one of the most applicable anaerobic treatments for dairy effluents. Fundamentally, the UASB reactor is composed of a sludge blanket, an influent-distribution system, a gas–solid separator, and an attached effluent-withdrawal port (Borja and Banks 1994). In a study, 97.5% COD removal and 98% lactose conversion were obtained using a conventional UASB reactor to treat dairy wastewater (Najafpour and Asadi 2008). In an investigation, the possibility of using UASB reactors was revealed for treating dairy wastewater to the magnitude of 96% COD reduction (Ramasamy et al. 2004).

An anaerobic filter (AF) reactor is proposed for the treatment of dairy effluents with low amounts of suspended solids. The basic function of the AF reactor is entrapment of suspended solids and giving enough retention time for biomaterials. Consequently, the retention time for suspended solids and hydraulic retention time (HRT) impacts the efficiency of the AF reactor. It needs to consider the porosity of supporting media, which considerably influences the efficiency of the AF reactor. Using the supporting media with a large surface area increases the attachment of biomaterials, and enhanced porosity results in the low required volume of the reactor and diminishes the clotting of the filter (Kushwaha et al. 2011). An investigation of the AF reactor for the treatment of industrial complex dairy wastewater achieved more than 90% COD reduction (Omil et al. 2003). In a pilot trial for dairy waste treatment using an up-flow AF reactor, an average COD removal of 70% was obtained (Monroy et al. 1994).

10.4 NON-THERMAL APPROACHES

10.4.1 Membrane Processing

Membrane processing refers to a group of size-based separation technologies that include microfiltration, ultrafiltration, nanofiltration, and reverse osmosis. These technologies are distinguished by the membrane type, pressure, and configuration. The pore size of the membranes is used to classify membrane processing. Table 10.3 shows the pore size range associated with each membrane technology.

Principles and characteristics of the membrane process used to treat WW are reviewed elsewhere (Chen et al. 2017; Catenacci et al. 2020). Membranes are coupled with biological reactors (MBR), offering some advantages – effluent quality and high removal efficiencies. An MBR coupled with a microfiltration membrane (pore size 0.5 μm) showed a removal efficiency of 99%, 89%, and 86% of COD, total nitrogen, and phosphorus, respectively, from UHT milk, yogurt, and

TABLE 10.3 PORE SIZE RANGE ASSOCIATED WITH MEMBRANE PROCESSING

Membrane Processing	Pore Size
Microfiltration	> 0.1 μm
Ultrafiltration	0.1–0.01 μm
Nanofiltration	0.01–0.001 μm
Reverse osmosis	< 0.001 μm

cheese WW (Andrade et al. 2013). An investigation of the treatment of dairy wastewater through RO resulted in 90–95% water recovery (Vourch et al. 2008). After treating dairy wastewater using single RO or NF + RO methods, the recovered water showed satisfactory properties such as total organic carbon (TOC) and conductivity for reusing for heating, cooling, and cleaning purposes. An investigation on the application of MBR coupled with nanofiltration as a tertiary treatment of WW showed the removal of 98% and 89% of the COD and phosphorus, respectively (Andrade et al. 2014).

10.4.2 Hydrodynamic Cavitation

The formation of cavities within a liquid, their growth, and subsequent collapse is known as cavitation (Osorio-Arias et al. 2021). In general, two kinds of cavitation are thought to occur when fluid is suddenly reduced below its vapor pressure (Ferrari 2017). The first type of cavitation is known as acoustic cavitation, and it relates to the propagation of waves through a region subjected to liquid vaporization. Hydrodynamic cavitation is the second type of cavitation, and it is the main effect in nozzles and discharge orifices (Sim et al. 2021). The cavitation dynamics depend on the flow conditions, fluid properties, surrounding pressure, and geometry of the valve. Over the past decade, significant progress has been made in the fundamental understanding of hydrodynamic cavitation.

Cavitation is based on the pressure–volume relationship of the van der Waals equation and the theory of probability of fluctuation (Endo 1994). During the transition from liquid to vapor, there is an interphase where both phases coexist, which corresponds to the length of the spinodal line of stability of the liquid. The probability of fluctuation theory predicts that cavitation frequency is proportional to the pressure fluctuation within the liquid. The formation of vapor bubbles and their subsequent collapse cause significant thermo-mechanical effects. Collapsing mechanisms of cavitation bubbles have been reviewed elsewhere (van Wijngaarden 2016). In general, the collapse of the cavitation bubbles occurs within a very short time span (milli- or microseconds), resulting in shockwaves that raise the liquid temperature. The violent collapse of the cavities generates free radicals such as OH– and H+ (Gogate et al. 2001), accelerating the hydrolysis of fatty acids (Suslick et al. 1997) and the breakdown of organic material. Indeed, hydrodynamic cavitation has been used to reduce the organic load in wastewater from the wood finishing and paper manufacturing industries (Badve et al. 2014). Hirooka et al. (2009) treated WW from milk parlors with nozzle-cavitation prior to the membrane bioreactor and found a reduction of 80% of the sludge compared to the control treatment.

10.4.3 Electrochemical Methods

In a broad sense, electrochemical methods are the processes and factors that affect the transport of charges throughout chemical phases. Charges are transported through the electrode by the movement of electrons. The essential reaction taking place is the reduction of metals from different oxidation states to the elemental state. Such reductions consist of electroflotation, electrooxidation, and electrocoagulation (Kushwaha et al. 2010).

10.4.3.1 Electroflotation

Electroflotation involves the formation of bubbles as a result of water electrolysis by means of anodic and cathodic reactions (Equations 1 and 2, respectively).

$$2H_2O \leftrightarrow O_2 + 4H^+ + 4e^- \qquad (1)$$

$$2H_2O + 2e^- \leftrightarrow H_2 + 2OH^- \qquad (2)$$

The size and number of formed bubbles depend on the composition and shape of the electrodes, applied current density, dissolved gasses in the stream, and temperature (Sillanpää and Shestakova 2017). Overall, the arrangement and configurations of the electroflotation reactors depend on the direction of water movement, and they are broadly classified as:

1) Countercurrent flow, where water and flotation gases move in the opposite direction.
2) Direct flow, which can be further divided into the horizontal or vertical arrangement of electrodes.
3) Mixed flow of water and gases.

Figure 10.1–10.3 illustrates the different configurations of the electroflotation reactors used to treat wastewater. These types of reactors employed electrodes made of copper or stainless steel, and they can be in the form of plates or wire mesh. A detailed description of the reactors used for electroflotation of WW can be found elsewhere (Sillanpää and Shestakova 2017). Overall, the electrodes are located at the bottom of the reactors and covered with the WW, allowing an even distribution of gas bubbles through the cross-section of the reactor. The WW enters the reactor through the inlet pipe located in the upper part, saturated with the formed gas bubbles. Upon bubbling, a froth layer is formed and subsequently removed from the reactor by the inclined channels. Inside the channels, a pipe with hot water contributes to the removal of the froth. The remaining sediments are discharged at the bottom of the reactor. The desired dispersion of gas bubbles can be obtained by changing the diameter of the wire (Figure 10.1). The arrangement of direct flow and mixed flow (Figures 10.2 and 10.3, respectively) are multicompartment, where the WW is delivered to the feeder compartment and separated from the main part of the reactor.

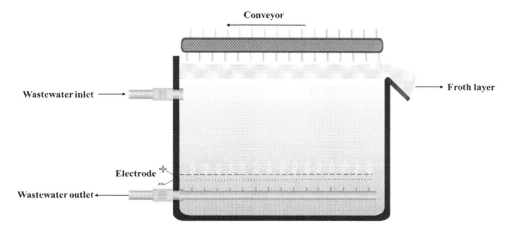

Figure 10.1 Electroflotation reactor with countercurrent arrangement. Source: reprinted from *Electrochemical Water Treatment Methods*, 2017, Mika Sillanpää and Marina Shestakova, with permission from Elsevier.

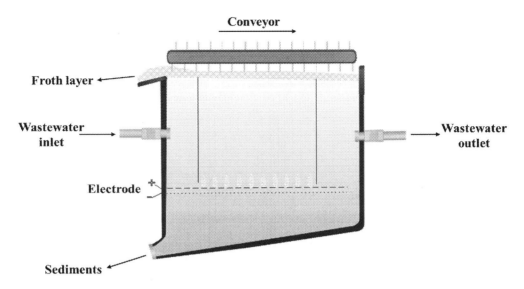

Figure 10.2 Electroflotation reactor with direct flow arrangement. Source: reprinted from *Electrochemical Water Treatment Methods*, 2017, Mika Sillanpää and Marina Shestakova, with permission from Elsevier.

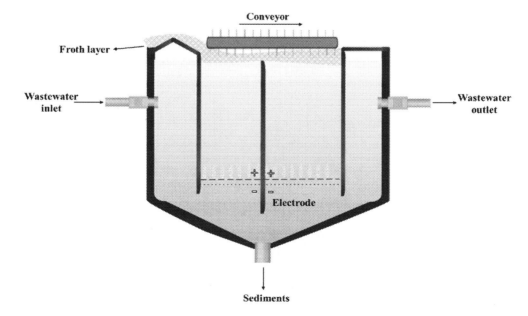

Figure 10.3 Electroflotation reactor with mixed flow. Source: reprinted from *Electrochemical Water Treatment Methods*, 2017, Mika Sillanpää and Marina Shestakova, with permission from Elsevier.

Once it enters the main compartment, the WW is saturated with gas bubbles generated at the electrode surface. Pollutants tend to form large flocs (froth layers) at the surface of the water, which are removed by mechanical devices (Sillanpää and Shestakova 2017).

10.4.3.2 Electrocoagulation

Electrocoagulation (EC) induces coagulation and sedimentation of pollutants in the WW. Overall, WW flows through a stack of electrodes where electric current is applied to form polynuclear complexes that interact with the pollutants within the WW stream, making them coagulate (Sillanpää and Shestakova 2017). Figures 10.4 and 10.5 show two types of arrangements for the stack of electrodes – horizontal and vertical, respectively. In the horizontal arrangement (Figure 10.4), the inlet pipe of WW is located in the upper part of the reactor.

At the end of the treatment, treated WW exits the reactor after flowing through the stack of electrodes, while the sediments are discharged at the bottom part of the reactor. For the vertical arrangement (Figure 10.5), WW enters the reactor from the bottom part and exists at the upper part. The stack of electrodes can be classified as single or multiflow modes, depending on the arrangement (Tahreen et al. 2020). In the single-flow mode, WW flows through a series of connected channels that provide sufficient residence time at moderate flow rates. On the other hand, the multiflow mode consists of the simultaneous flow of WW through multiple channels equipped with electrodes. Electrocoagulation reactors have been integrated with filtration or sedimentation to clarify the water in one step. Characteristics and arrangements of such reactors can be found elsewhere (Sillanpää and Shestakova 2017).

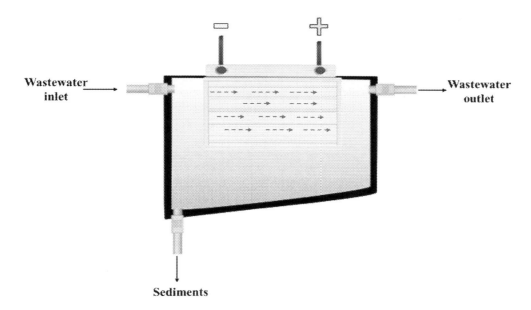

Figure 10.4 Electrocoagulaton reactor with horizontal arrangement. Source: reprinted from *Electrochemical Water Treatment Methods*, 2017, Mika Sillanpää and Marina Shestakova, with permission from Elsevier.

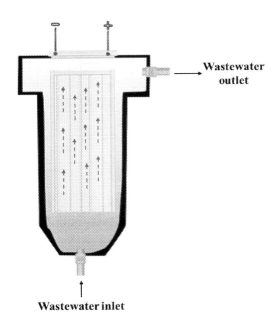

Figure 10.5 Electrocoagulaton reactor with vertical arrangement. Source: reprinted from *Electrochemical Water Treatment Methods*, 2017, Mika Sillanpää and Marina Shestakova, with permission from Elsevier.

TABLE 10.4 ADVANCED OXIDATION METHODS FOR THE TREATMENT OF WASTEWATER

Entry	Technology
1	Ozone
2	Ozone/UV
3	Ozone/H_2O_2
4	Ozone/TiO_2
5	H_2O_2/Fe^{2+} (Fenton)
6	Ozone/ultrasound

10.4.4 Advanced Oxidation Methods

Advanced oxidation methods or processes (AOPs) are based on the generation of hydroxyl free radicals. Table 10.4 summarizes the different ways to generate hydroxyl radicals. Fundamentals and applications of the different advanced oxidation methods can be found elsewhere (Bajpai 2017).

Overall, AOPs generate chemical radicals having one unpaired electron, which further triggers a set of oxidation reactions (Bijan and Mohseni 2005). The oxidation process stops with the formation of CO_2 and H_2O. The way in which the radicals are formed can be nonphotochemical and photochemical. In the first case, the radicals are initiated when oxidizing agents (ozone and hydrogen peroxide) are applied into the water. The photochemical process uses a combination of oxidizing agents with UV irradiation. A number of mechanisms are involved during the reactions between hydroxyl radicals and organic compounds, depending on the nature of the substrate (Bajpai 2017). The mechanism by which the hydroxyl radicals react with organic compounds includes an abstraction of hydrogen from an adjacent molecule, a pollutant, and the unpaired electron transfer to form carbonates (Munter et al. 1993). Thus, three levels of degradation are obtained (Silva and Jardim 2006):

1) Primary degradation, where structural changes occurr in the compounds.
2) Acceptable degradation, where decomposition of the compounds reduces their toxicity.
3) Final degradation, which involves the mineralization of organic compounds.

10.4.5 Subcritical Water

Subcritical water has recently attracted researchers due to its unique characteristics as a reaction medium (Enteshari and Martínez-Monteagudo 2018). This technology is also known as subcritical water hydrolysis, water above the boiling and below the critical point (Möller et al. 2011). The rationality of using pressurized water lies in physicochemical properties induced by the combination of pressure and temperature. Thermodynamic properties of water under pressure are well documented and available through the National Institute of Standards and Technology database. In summary, water within the subcritical state possesses relatively low dielectric permittivity and high ion strength, which favors reactions such as nucleophilic substitution, eliminations, and hydrolysis (Brunner 2009). Hydrolysis as a result of subcritical water offers unique advantages in valorizing wastewater streams. For instance, the waste stream's water content acts as a reaction agent and solvent, meaning that streams of wastewater can be directly treated without any pretreatment. The hydrolysis occurring in subcritical water is often accompanied by secondary reactions (oxidation, hemolysis, and additions). These reactions are mainly driven by the temperature and might influence the final products of the hydrolysis treatment.

The use of subcritical water to produce peptides from ice cream WW has been illustrated (Enteshari and Martínez-Monteagudo 2018). These authors produced peptides with desirable functionality, including antioxidant and antihypertensive capacity as well as improved solubility. Ice cream wastewater collected from the return line of a typical clean-in-place protocol was hydrolyzed within the subcritical water domain (130–230° C and 15–60 bar). Figure 10.6 shows the evolution of the degree of hydrolysis at different temperatures. Such experiments allow us to identify reaction conditions yielding the maximum degree of hydrolysis.

More importantly, the subcritical hydrolysis was mathematically represented using a probabilistic model from which the conditions needed to obtain a given degree of hydrolysis can be illustrated (Figure 10.7). The application of the Weibull model for subcritical hydrolysis assumes that the reaction followed a probabilistic exponential distribution, regardless of the reaction mechanism. After subcritical hydrolysis, the hydrolysates were recovered, and their antioxidant and antihypertensive capacity were determined (Figure 10.7). It was found that there was a strong linear relationship between antioxidant capacity and the degree of hydrolysis (Figure 10.7a). Similarly, the antihypertensive capacity was shown to be influenced by the degree of hydrolysis (Figure 10.7b). Subcritical hydrolysis offers the opportunity to produce biologically active hydrolysate, and the combined action of different fractions may lead to a greater antioxidant and antihypertensive capability. Nevertheless, further research is certainly necessary to link the inhibitory ability of the recovered fraction with a specific group of peptides. Additionally, the amino acid profile before and after subcritical hydrolysis was considerably different, especially hydroxyproline and glutamic acid.

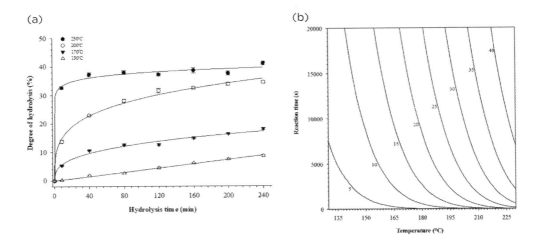

Figure 10.6 (a) Kinetics of subcritical hydrolysis of ice cream wastewater. (b) Temperature–time combinations needed to achieve a specific level of hydrolysis. Source: reprinted from "Subcritical hydrolysis of ice-cream wastewater: Modeling and functional properties of hydrolysate," *Food and Bioproducts Processing* 111, 2018, Maryam Enteshari and Sergio I. Martinez-Monteagudo, with permission from Elsevier.

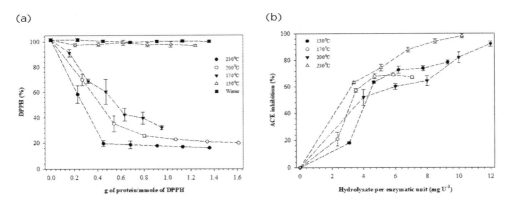

Figure 10.7 (a) Antiradical ability of the hydrolysate recovered after subcritical hydrolysis of ice cream wastewater. (b) Inhibitory ability of angiotensin I-converting enzyme (ACE) of the hydrolysate recovered after subcritical hydrolysis of ice cream wastewater. Source: reprinted from "Subcritical hydrolysis of ice-cream wastewater: Modeling and functional properties of hydrolysate," *Food and Bioproducts Processing* 111, 2018, Maryam Enteshari and Sergio I. Martinez-Monteagudo, with permission from Elsevier.

10.5 CONCLUSIONS

Many emerging technologies have been used to treat dairy WW, including membrane filtration, ultrasound, hydrodynamic cavitation, subcritical water, and advanced oxidation methods. Each technology offers unique features, and coupling non-thermal technologies with biological treatments may be a convenient solution for integrating the benefits of both technologies. Future research trends in relation to non-thermal technologies may focus on forming and recovering value-added compounds, such as peptides, fatty acids, carbohydrates, and micronutrients.

REFERENCES

Ahmad, T., R. M. Aadil, H. Ahmed, Uu Rahman, B. C. V. Soares, S. L. Q. Souza, T. C. Pimentel, H. Scudino, J. T. Guimarães, E. A. Esmerino, M. Q. Freitas, R. B. Almada, S. M. R. Vendramel, M. C. Silva, and A. G. Cruz. 2019. "Treatment and utilization of dairy industrial waste: A review." *Trends in Food Science and Technology* 88:361–372. doi: 10.1016/j.tifs.2019.04.003.

Andrade, L. H., F. D. S. Mendes, J. C. Espindola, and M. C. S. Amaral. 2014. "Nanofiltration as tertiary treatment for the reuse of dairy wastewater treated by membrane bioreactor." *Separation and Purification Technology* 126:21–29. doi: 10.1016/j.seppur.2014.01.056.

Andrade, L. H., G. E. Motta, and M. C. S. Amaral. 2013. "Treatment of dairy wastewater with a membrane bioreactor." *Brazilian Journal of Chemical Engineering* 30(4):759–770.

Badve, Mandar P., Parag R. Gogate, Aniruddha B. Pandit, and Levente Csoka. 2014. "Hydrodynamic cavitation as a novel approach for delignification of wheat straw for paper manufacturing." *Ultrasonics Sonochemistry* 21(1):162–168. doi: 10.1016/j.ultsonch.2013.07.006.

Bajpai, Pratima. 2017. "Chapter 7.Emerging technologies for wastewater treatment." In *Pulp and Paper Industry*, edited by Pratima Bajpai, 93–179. Elsevier. ISBN 9780128110997.

Bijan, Leila, and Madjid Mohseni. 2005. "Integrated ozone and biotreatment of pulp mill effluent and changes in biodegradability and molecular weight distribution of organic compounds." *Water Research* 39(16):3763–3772. doi: 10.1016/j.watres.2005.07.018.

Borja, R., and C. J. Banks. 1994. "Kinetics of an upflow anaerobic sludge blanket reactor treating ice-cream wastewater." *Environmental Technology* 15(3):219–232. doi: 10.1080/09593339409385423.

Brunner, G. 2009. "Near critical and supercritical water. Part I. Hydrolytic and hydrothermal processes." *The Journal of Supercritical Fluids* 47(3):373–381. doi: 10.1016/j.supflu.2008.09.002.

Carvalho, Fátima, Ana R. Prazeres, and Javier Rivas. 2013. "Cheese whey wastewater: Characterization and treatment." *Science of the Total Environment* 445–446:385–396. doi: 10.1016/j.scitotenv.2012.12.038.

Catenacci, Arianna, Micol Bellucci, Tugui Yuan, and Francesca Malpei. 2020. "Chapter 9.Dairy wastewater treatment using composite membranes." In *Current Trends and Future Developments on (Bio-) Membranes*, edited by Angelo Basile and Antonio Comite, 261–288. Elsevier. ISBN 978-0-12-816823-3.

Chen, Zhiwei, Jianquan Luo, Yujue Wang, Weifeng Cao, Benkun Qi, and Yinhua Wan. 2017. "A novel membrane-based integrated process for fractionation and reclamation of dairy wastewater." *Chemical Engineering Journal* 313:1061–1070. doi: 10.1016/j.cej.2016.10.134.

Danalewich, J. R., T. G. Papagiannis, R. L. Belyea, M. E. Tumbleson, and L. Raskin. 1998. "Characterization of dairy waste streams, current treatment practices, and potential for biological nutrient removal." *Water Research* 32(12):3555–3568. doi: 10.1016/S0043-1354(98)00160-2.

Endo, Harumi. 1994. "Thermodynamic consideration of the cavitation mechanism in homogeneous liquids." *The Journal of the Acoustical Society of America* 95(5):2409–2415. doi: 10.1121/1.409850.

Enteshari, Maryam, and Sergio I. Martínez-Monteagudo. 2018. "Subcritical hydrolysis of ice-cream wastewater: Modeling and functional properties of hydrolysate." *Food and Bioproducts Processing* 111:104–113. doi: 10.1016/j.fbp.2018.08.002.

Enteshari, Maryam, and Sergio I. Martínez-Monteagudo. 2020. "Hydrothermal conversion of ice-cream wastewater." *Journal of Food Process Engineering* 43(10):e13498. doi: 10.1111/jfpe.13498.

Ferrari, A. 2017. "Fluid dynamics of acoustic and hydrodynamic cavitation in hydraulic power systems." *Proceedings. Mathematical, Physical, and Engineering Sciences* 473(2199):20160345. doi: 10.1098/rspa.2016.0345.

Gavala, H. N., H. Kopsinis, I. V. Skiadas, K. Stamatelatou, and G. Lyberatos. 1999. "Treatment of dairy wastewater using an upflow anaerobic sludge blanket reactor." *Journal of Agricultural Engineering Research* 73(1):59–63. doi: 10.1006/jaer.1998.0391.

Gogate, Parag R., Irfan Z. Shirgaonkar, M. Sivakumar, P. Senthilkumar, Nilesh P. Vichare, and Aniruddha B. Pandit. 2001. "Cavitation reactors: Efficiency assessment using a model reaction." *AIChE Journal* 47(11):2526–2538. doi: 10.1002/aic.690471115.

Hirooka, Kayako, Ryoki Asano, Atsushi Yokoyama, Masao Okazaki, Akira Sakamoto, and Yutaka Nakai. 2009. "Reduction in excess sludge production in a dairy wastewater treatment plant via nozzle-cavitation treatment: Case study of an on-farm wastewater treatment plant." *Bioresource Technology* 100(12):3161–3166. doi: 10.1016/j.biortech.2009.01.011.

Janczukowicz, W., M. Zieliński, and M. Dębowski. 2008. "Biodegradability evaluation of dairy effluents originated in selected sections of dairy production." *Bioresource Technology* 99(10):4199–4205. doi: 10.1016/j.biortech.2007.08.077.

Karadag, D., O. E. Köroğlu, B. Ozkaya, and M. Cakmakci. 2015. "A review on anaerobic biofilm reactors for the treatment of dairy industry wastewater." *Process Biochemistry* 50(2):262–271. doi: 10.1016/j.procbio.2014.11.005.

Khalaf, Alaa H., W. A. Ibrahim, Mai Fayed, and M. G. Eloffy. 2021. "Comparison between the performance of activated sludge and sequence batch reactor systems for dairy wastewater treatment under different operating conditions." *Alexandria Engineering Journal* 60(1):1433–1445. doi: 10.1016/j.aej.2020.10.062.

Kushwaha, J. P., V. C. Srivastava, and I. D. Mall. 2010. "Organics removal from dairy wastewater by electrochemical treatment and residue disposal." *Separation and Purification Technology* 76(2):198–205. doi: 10.1016/j.seppur.2010.10.008.

Kushwaha, J. P., V. C. Srivastava, and I. D. Mall. 2011. "An overview of various technologies for the treatment of dairy wastewaters." *Critical Reviews in Food Science and Nutrition* 51(5):442–452. doi: 10.1080/10408391003663879.

Möller Maria, Peter Nilges, Falk Harnisch, and Uwe Schröder. 2011. "Subcritical water as reaction environment: Fundamentals of hydrothermal biomass transformation." *ChemSusChem* 4(5):566–579. doi: 10.1002/cssc.201000341.

Monroy, O., K. A. Johnson, A. D. Wheatley, F. Hawkes, and M. Caine. 1994. "The anaerobic filtration of dairy waste: Results of a pilot trial." *Bioresource Technology* 50(3):243–251. doi: 10.1016/0960-8524(94)90097-3.

Munter, R., S. Preis, S. Kamenev, and E. Siirde. 1993. "Methodology of ozone introduction Into water and wastewater treatment." *Ozone: Science and Engineering* 15(2):149–165. doi: 10.1080/01919519308552265.

Najafpour, G., and M. Asadi. 2008. "Biological treatment of dairy wastewater in an upflow anaerobic sludge-fixed film bioreactor." *American-Eurasian Journal of Agricultural and Environmental Science* 4:251–257.

Omil, Francisco, Juan M. Garrido, Belén Arrojo, and Ramón Méndez. 2003. "Anaerobic filter reactor performance for the treatment of complex dairy wastewater at industrial scale." *Water Research* 37(17):4099–4108. doi: 10.1016/S0043-1354(03)00346-4.

Osorio-Arias, Juan C., Oscar Vega-Castro, and Sergio I. Martínez-Monteagudo. 2021. "3.15.Fundamentals of high-pressure homogenization of foods." In: *Innovative Food Processing Technologies*, edited by Kai Knoerzer and Kasiviswanathan Muthukumarappan, 244–273. Oxford: Elsevier.

Ramasamy, E. V., S. Gajalakshmi, R. Sanjeevi, M. N. Jithesh, and S. A. Abbasi. 2004. "Feasibility studies on the treatment of dairy wastewaters with upflow anaerobic sludge blanket reactors." *Bioresource Technology* 93(2):209–212. doi: 10.1016/j.biortech.2003.11.001.

Rusten, B., A. Lundar, O. Eide, and H. Ødegaard. 1993. "Chemical pretreatment of dairy wastewater." *Water Science and Technology* 28(2):67–76. doi: 10.2166/wst.1993.0078.

Schwarzenbeck, N., J. M. Borges, and P. A. Wilderer. 2005. "Treatment of dairy effluents in an aerobic granular sludge sequencing batch reactor." *Applied Microbiology and Biotechnology* 66(6):711–718. doi: 10.1007/s00253-004-1748-6.

Sillanpää, Mika, and Marina Shestakova. 2017. "Chapter 4.Equipment for electrochemical water treatment." In *Electrochemical Water Treatment Methods*, edited by Mika Sillanpää and Marina Shestakova, 227–263. Butterworth-Heinemann. ISBN 9780128131602.

Silva, Leonardo M. da, and Wilson F. Jardim. 2006. "Trends and strategies of ozone application in environmental problems." *Química Nova* 29(2):310–317.

Sim, Jae Young, Steven L. Beckman, Sanjeev Anand, and Sergio I. Martínez-Monteagudo. 2021. "Hydrodynamic cavitation coupled with thermal treatment for reducing counts of B. coagulans in skim milk concentrate." *Journal of Food Engineering* 293. doi: 10.1016/j.jfoodeng.2020.110382. 110382.

Suslick, Kenneth S., Millan M. Mdleleni, and Jeffrey T. Ries. 1997. "Chemistry induced by hydrodynamic cavitation." *Journal of the American Chemical Society* 119(39):9303–9304. doi: 10.1021/ja972171i.

Tahreen, Amina, Mohammed Saedi Jami, and Fathilah Ali. 2020. "Role of electrocoagulation in wastewater treatment: A developmental review." *Journal of Water Process Engineering* 37. doi: 10.1016/j.jwpe.2020.101440. 101440.

van Wijngaarden, Leen. 2016. "Mechanics of collapsing cavitation bubbles." *Ultrasonics Sonochemistry* 29:524–527. doi: 10.1016/j.ultsonch.2015.04.006.

Vourch, M., B. Balannec, B. Chaufer, and G. Dorange'. 2008. "Treatment of dairy industry wastewater by reverse osmosis for water reuse." *Desalination* 219(1–3):190–202. doi: 10.1016/j.desal.2007.05.013.

Wang, Yifei, and Luca Serventi. 2019. "Sustainability of dairy and soy processing: A review on wastewater recycling." *Journal of Cleaner Production* 237. doi: 10.1016/j.jclepro.2019.117821. 117821.

Chapter 11

Surface Pasteurization and Disinfection of Dairy Processing Equipment Using Cold Plasma Techniques

Kamalapreetha Baskaran and Mahendran Radhakrishnan

CONTENTS

11.1	Introduction	143
11.2	Mechanism of Cold Plasma and Its Interaction with the Surface	145
11.3	Current Methods of Surface Pasteurization and Disinfection in the Dairy Industry	146
11.4	Effect of Cold Plasma on Food Contact Surfaces	147
11.5	Effects of Plasma-Activated Water in Cleaning and Sanitization	148
11.6	Scope and Future Trends of Cold Plasma Techniques in Dairy Processing Equipment	151
11.7	Conclusion	153
References		153

11.1 INTRODUCTION

Milk is a perishable food product that is susceptible to contamination and spoilage. In the dairy processing line, equipment surfaces are regarded as the critical source of biological contamination of the processed milk. Adhered contaminants reduce the quality and shelf-life of milk. In this regard, the dairy industry needs to decontaminate the milk processing line before every pass. Disinfestation and surface pasteurization are the most significant steps in the dairy production line. Various cleaning and sanitation methods and procedures are being implemented to clean equipment, mainly involving transportation, processing, handling, and storage of milk or milk products. Developed biofilms/stains on the surfaces of food production lines are tough to eliminate and prevent from recontamination even after applying an adequate disinfecting agent or following a cleaning procedure. This chapter focuses on applying emerging cold plasma technology as an alternative approach to pasteurization and disinfection of dairy processing equipment.

Cold plasma is a non-thermal processing technology that simultaneously utilizes various reactive species (hurdle concept) to act on the target of interest. Plasma cleaning is a process employing a non-equilibrated, ionizing gas called plasma, which removes all organic matter from the surface of an object. Plasma is produced by different means of supplying energy to gas

(Mir et al., 2020). It is usually achieved using oxygen and/or argon gas. Cleaning with plasma is an eco-friendly process where toxic chemicals do not prevail. During and after the plasma process, Scholtz et al. (2015), Ratish Ramanan et al. (2018), and Paatre et al. (2020) detected the generation of various reactive species like free electrons and radicals, positive and negative ions, neutral and excited atoms, molecules, and ultraviolet (UV) photons. Among the produced species, reactive oxygen (ROS) and reactive nitrogen (RNS) are the significant components responsible for the detrimental effect on pathogenic microbes (Iuchi et al., 2018). Plasma exhibited free radical or reactive species, which clean the equipment's surface and increase the adherence nature of the surface. Mahendran and Alagusundaram (2015) have recommended that cold plasma might have an increased advantage as an effective method against surface contamination in food processing industries. Sainz-García et al. (2021) reported that even at atmospheric pressure and ambient temperature, plasma is generated with compressed air and electricity. Besides, atmospheric pressure cold plasma is inexpensive technology, toxin-free, and does not involve any supporting system. Some of the methods of plasma generation include:

Corona discharge plasma: It involves the generation of plasma near the sharp edges. Tremendous energy is supplied to surfaces with two different radii in curvature; one is a sharp edge called the emitter, and the other can be a thin wire that acts as a collector (Feizollahi, Misra, and Roopesh, 2021). Corona discharge plasma is more prominent in the production of ozone generation for commercial purposes. The electric field between the two electrodes is maintained at a minimum range to prevent spark discharge.

Dielectric barrier discharge (DBD) plasma: DBD plasma is the most widely used plasma with two electrodes of similar sizes, placed parallel and coated with a dielectric material. The simple, flexible geometrical design, easy construction, and operating parameters make it versatile compared to other plasma discharges (Sakudo, Misawa, and Yagyu, 2019). The major advantage of DBD plasma is that it aids in reducing the formation of spark discharges and produces streamer discharge. DBD operation in a low-pressure environment with changes in the combinations of electrode distance and pressure values generates a glow discharge plasma (Turner, 2016). Here, a plasma plume is produced between the two electrodes with variations in the layers of light emission patterns between electrodes.

Resistive barrier discharge plasma: In this kind of discharge, the dielectric barrier in the DBD plasma is replaced by a resistive barrier. The resistive barrier maintains the streamer discharge from changing into spark discharge by limiting current to the circuit using a resistive impedance (Pedrow et al., 2020).

Microwave discharge plasma: Electromagnetic radiations of 300 MHz to 300 GHz frequency is used for generating plasma. This kind of discharge is also called electrodeless discharge; it can be generated using microwave waveguides (Feizollahi, Misra, and Roopesh, 2021).

Plasma jet discharge: Plasma jets are atmospheric pressure plasma provided with non-sealed electrodes and a supply of gases for jet formation. Plasma jets can be made of either DBD or corona discharge, where the supply of gas carries the plasma produced between the electrodes. This leads to the projection of a plasma plume outside the device and into the environment (Sousa et al., 2011). Plasma jets are mainly used for direct application, provided that the target surface is not a limitation in this type of discharge.

Based on the need for sterilizing the processing equipment surfaces, the size of the equipment must not be a limitation. Thus, among all the kinds of discharges, the best suitable discharge for the application of plasma on the equipment surface is the plasma jet. Apart from these types of plasma discharges, plasma can also be subjected to water to activate it (Chen et al., 2019). The activated liquid (with dissolved reactive species), which is loaded with antimicrobial properties, can also be used for cleaning processing equipment.

11.2 MECHANISM OF COLD PLASMA AND ITS INTERACTION WITH THE SURFACE

There are five concepts involved in the mechanism of cold plasma interaction: ablation, chemical etching, activation, deposition, and crosslinking. The mechanism of cold plasma is depicted in Figure 11.1.

Plasma ablation: Includes the mechanical evacuation of contaminants present on the surface by ion bombardment and energetic electrons. The outside substrate material molecular layer, as well as contaminant layers, are only affected by this plasma ablation. Because of its chemical inertness and high ablation efficiency with the surface material, argon gas is widely used (Keidar et al., 2011).

Chemical etching: Chemical etching includes the chemical response of films or surface organic toxins with profoundly reactive free radicals in cold plasma to give out volatile byproducts that are released from the surface of the sample. By legitimate choice of the gas chemistry, different sorts of materials can be etched specifically while limiting the etching of various materials on the surface of the sample. Oxygen is mostly used to etch chemically and also to eliminate organic materials from the surface (Thirumdas, Kadam, and Annapure, 2017).

Activation: Plasma surface activation includes the formation of surface chemical functional groups using plasma gases like hydrogen, oxygen, ammonia, and nitrogen, which separate and respond with the surface. In the circumstance of polymers, weaker surface bonds are broken by plasma and replaced by chemical groups like hydroxyl,

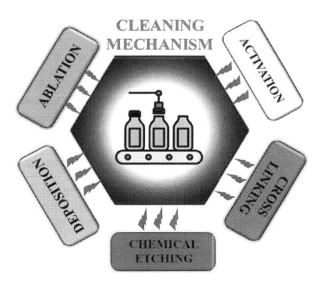

Figure 11.1 Plasma cleaning mechanism.

carboxyl, or carbonyl groups. This actuation modifies the wettability of the surface and chemistry, which can extraordinarily increase attaching and gripping to other surfaces (Liao et al., 2018).

Deposition: Plasma deposition includes the arrangement of thin coated polymer on the surface through polymerization of the interaction gas. The process gas is a combination of inert gas and vaporized monomer to be utilized for plasma generation. With the presence of a single chemical functional group, deposition of the polymer layer can occur by applying plasma polymerization (Los et al., 2019).

Crosslinking: Crosslinking is the linkage of the atomic chain and bonding in a polymer. Inert gas with plasma can be utilized with crosslink polymers and produce a more grounded and stronger substrate microsurface. Under particular conditions, chemical resistance or additional wear to a material can occur when crosslinking through plasma is done (Pankaj et al., 2014).

11.3 CURRENT METHODS OF SURFACE PASTEURIZATION AND DISINFECTION IN THE DAIRY INDUSTRY

Disassembling the processing line for cleaning after every pass is a tedious process and expensive too. In a food processing unit, the production line must accomplish hygienic conditions. It should be devoid of contamination and deterioration of the product, and cleaning of the plant must be carried out quickly and thoroughly. Currently, cleaning criteria are met in the food processing plant by the cleaning-in-place (CIP) systems. CIP systems provide quick, effective, and reliable cleaning of all forms of process plants. It is a system that cleans full plant equipment or pipeline circuits without dismantling the equipment. Memisi et al. (2015) discussed the CIP system of the dairy industry. There are two types of clean-in-place (CIP) systems in the dairy industry: closed system (lost washing) and automatic system (reusing washed solution). CIP has various advantages of superior operational safety and quality sanitization and is operated at a controlled cost using resources (sanitation, water, and energy). CIP cleaning consists of five steps of washing using cold water, an acid, a base, steamy water, and hot water. This cleaning solution is allowed to pass into equipment only through pipes, heaters, pumps, valves, and separators at a turbulent flow. These washing steps form friction, which ultimately leads to the removal of the surface deposits.

Jiménez-Pichardo et al. (2016) studied the alternative method for traditional cleaning and disinfection procedure on stainless-steel plates with and without polishing using the combined effect of neutral (NEW) and alkaline (AEW) electrolyzed water. This method can effectively be useful in the dairy plant for surface sanitization with reduced temperature and cost. This study observed a significant removal of adhered microorganisms on the surface of the equipment. There is no significant difference in the reduction of microbial load in polished (3.20 log CFU/cm^2) and non-polished surface (3.90 log CFU/cm^2) at 30° C, where the treatment time is for 30 min and 10 min at a concentration of AEW of 300 mg/L and 100 mg/L for polished and non-polished respectively. The most common solution for decontamination on a food contact surface is the chlorine solution. Greene, Few, and Serafini (1993) investigated the inactivation of *Pseudomonas fluorescens* and *Alcaligenes faecalis*, bacteria responsible for milk spoilage, using ozonated and chlorinated water on a stainless-steel surface. Chlorine is applied at 100 ppm for 120 s and ozonated deionized water is treated at 0.5 ppm for 10 min on *P. fluorescens* and *A. faecalis*. Both the ozonation and chlorination are considerable anti-biofilms treatments with > 99.99% inactivation efficiency.

Witte et al. (2017) reported and investigated the cleaning efficiency of dairy equipment with dry ice (solid CO_2) as another method. Dry ice blasting is an eco-friendly, fast, and residual-free method with a slightly abrasive effect. However, it is considered to be effective in cleaning but not as a disinfecting agent. The amount of dry ice used and the pressure variation significantly affect the treatment's effectiveness rather than the pellet's size or the microbes' initial load on the surface. Dry ice blasting is a great alternative for cleaning but with due care to be taken as it has a high risk of recontamination of microbes.

Disinfectants like GALA for dishes, FAIRY Juicy Lemon, as well as alkaline agents and acid agents were used in the cleaning of milking machines. Zhukorskyi and Kryvokhyzha (2016) state that chemical sanitizing agents' application observed chlorine compounds, anion surface-active substances, phosphates, silicates, cation SAS, non-ionogenic SAS, and acids in the surrounding environment within one year of treatment, which led to a weakening of the natural biogeocenoses.

In dairy industries, *Staphylococcus aureus* and *S. enterica* are the most toxic microbes present and develop a biofilm on the surface, which will affect the quality of the raw milk. Bayoumi et al. (2012) examined the NaOCl effect on the stainless surface and polypropylene surface to destroy adherent biofilm. Varga and Szigeti (2016) emphasized the oxidizing effect of toxic ozone gas on the microbial quality of the developed biofilm, which resulted in a detrimental effect on the mold growth on stainless-steel surfaces of the dairy industry.

11.4 EFFECT OF COLD PLASMA ON FOOD CONTACT SURFACES

In Sainz-García et al. (2021), atmospheric pressure cold plasma (APCP) technology is utilized to sanitize oak wood staves and their impact on the re-utilization of barrels. Three wine spoilage microorganisms (*P. pentosaceus*, *A. pasteurianus*, and *B. bruxellensis*) were treated at two different plasma powers (500 W for air and 90 W for argon) and three plasma gases (air, oxygen, and nitrogen). For APCP treatment, *B. bruxellensis* observed to be the most sensitive. Air and nitrogen gas plasma show a higher efficacy in terms of microbial inactivation because of the reactive species produced. There were no structural changes on the wood surface observed during the APCP treatments, which leads the way toward the reuse of the barrels for further processing, which is also free from chemical and microbial contamination.

Gabriel et al. (2016) studied the effect of an atmospheric pressure plasma jet on four developed strains of *Pseudomonas aeruginosa* biofilm on the stainless-steel types 316 and 304 at three different finishes, namely, the mirror (MR), hairline (HL), and 2B surfaces. The surface finish of the stainless steel greatly impacted the biofilm formation. Microbial destruction or inactivation was observed due to the reactive species, UV-C radiation, and instant temperature rise. This study suggested that atmospheric pressure plasma can be used in industrial food sanitization and prevent the cross-contamination of *Pseudomonas* species on the food contact surface. It is an alternative approach with the benefits of being non-toxic and non-thermal in nature.

In a study by Niemira, Boyd, and Sites (2014), cold plasma rapid sanitization was used to prevent the cross-contamination of foods from pathogenic *Salmonella* biofilms. A three-strain *Salmonella* culture was allowed to develop a biofilm on a glass plate, and the treatment was carried out by placing it on a conveyor belt under varied voltage frequency and exposure time using filtered gas. Increasing the distance decreased the effectiveness of the treatment. At a shorter treatment time, cold plasma treatment on model food contact material attained a 2.13 log reduction in *Salmonella*. Cold plasma can be utilized as a rapid cleanser solution for food processing applications. Sarmiento et al. (2012) studied the washing efficiency of low-pressure plasma (oxygen and argon) on a steel sheet infused with a mixture of SAE 40 oil and diesel. According to the chromatogram, 72% cleaning efficiency was achieved on the metal surface and the treatment subsequently reduced wastewater production after pre-cleaning with a solvent (CH_3OH) followed by oxygen plasma.

In a study by Park and Ha (2018), at a higher treatment time (300 s), cold oxygen plasma showed a greater sensitivity toward the foodborne pathogen MNV-1 (norovirus – 3.89 log units) than HAV (2.02 log units) on the stainless-steel surface. Muranyi, Wunderlich, and Heise (2008) examined the effect of cascaded dielectric barrier discharge (CDBD) on a PET surface infused with *Aspergillus niger* conidiospores and *Bacillus subtilis* at varied relative humidity. Application of cold plasma technology on conidiospores and *Bacillus subtilis* spores achieved 2.6 and 5 log reduction, respectively. Dasan et al. (2017) inoculated *Escherichia coli* and *Staphylococcus epidermidis* and studied the decontamination effect of gliding arc discharge plasma on stainless steel (SS), silicone (Si), and PET surfaces. For 5 min of treatment, *E. coli* on SS, Si, and PET showed significant reductions of 2.72, 4.43, and 3.18 log. After 119 s of treatment, *Staphylococcus epidermis* showed inactivations of 3.76, 3.19, and 2.95 log on SS, Si, and PET surfaces respectively. All the treatments were found to maintain and operate at a temperature below 35° C.

The use of cold plasma for sanitation has far-ranging applications. Such applications were studied by Wang et al. (2003) on *Listeria monocytogenes* attached to stainless steel surfaces. The deactivation of *L.monocytogenes* by radio-frequency plasma was attained when modifying the stainless-steel surface with polyglycol-like structures under diethylene glycol vinyl ether (DIEGVE). The DIEGVE cold plasma can be a promising technique for reducing bacterial contamination and has great potential for biofilm inactivation.

Various studies have been made on dairy spoiling microbes' decontamination, cleaning, and sanitization using cold plasma on the food contact surface. The effect of plasma sanitation on some of the food contact surfaces is represented in Table 11.1. Cold plasma has a great impact on the degradation of spoilage microbial biofilms on the food contact surface. Since various research supports the application of cold plasma as a novel technique for the sterilization of processing equipment with respect to the plasma generating system and various sources of gases, cold plasma can also be applied as a sterilization agent on dairy processing equipment against disease-causing and contaminating spoilage biofilms. Still, research has to be done on the different finishes of the food contact surface and dairy spoilage–causing microbes.

11.5 EFFECTS OF PLASMA-ACTIVATED WATER IN CLEANING AND SANITIZATION

Exposing pure water to non-thermal plasma is called plasma-activated water (PAW). PAW has an antimicrobial effect because of the presence of reactive species generated by the plasma generation. Direct plasma treatment is limited to a shorter time and has a lesser lifetime. In contrast, PAW, consisting of stable particles such as nitrogen oxides (NOx) and corresponding acids, hydrogen peroxide (H_2O_2), and ozone (O_3), has a longer lifetime of decontamination effect. Cullen et al. (2018) suggested the effect of plasma technology, which aims at targeting the microbes using created species from plasma discharge on water and other liquids. Plasma-treated water

11.5 Effects of Plasma-Activated Water in Cleaning and Sanitization 149

TABLE 11.1 EFFECT OF PLASMA SANITIZATION ON CONTACT SURFACES

Contact Surface	Target Group	Types of Plasma	Effects	Reference
Stainless steel (SS) 304 and 316	*Pseudomonas aeruginosa* biofilm	Atmospheric pressure jet plasma	D value: SS 304 – 1.95 to 3.27 s and SS 316 – 2.53 to 3.16 s. Temperature raised in the range of 143.42 to 174.05° C within 15 s of treatment	Gabriel et al. (2016)
Steel sheet	Oil	Radiofrequency low-pressure plasma	The efficiency of cleaning increased to 72% with atmospheric oxygen low-pressure plasma	Sarmiento et al. (2012)
Glass slides	*Salmonella* strain	Plasma jet emitter	2.13 log destruction at a shorter exposure time of 15 s	Niemira, Boyd, and Sites (2014)
Stainless steel	Murine norovirus (MNV-1) and Hepatitis A virus (HAV)	—	At higher treatment time (300 s), cold oxygen plasma showed a greater sensitivity toward MNV-1 (3.89 log units) than HAV (2.02 log units)	Park and Ha (2018)
PET surface	*Aspergillus niger* and *Bacillus subtilis*	Dielectric barrier discharge plasma	At higher relative humidity (70%), maximum reduction of *A. niger* was observed as 2.6 log	Muranyi, Wunderlich, and Heise (2008)
Stainless steel 316	*Enterobacter sakazakii*	Radiofrequency plasma polymerization	Steel surface modified using ethylenediamine (EDA) shows a reduction of 99.74% at 45 W and 10 min	Sen et al. (2012)
Stainless steel, Silicon, and polyethylene terephthalate	*Escherichia coli* and *Staphylococcus epidermidis*	Gliding arc discharge	Nitrogen plasma shows a significant reduction in artificially impregnated microbes on the surfaces which operated at a lower temperature (35° C) at 5 min of exposure	Dasan et al. (2017)
Modified stainless steel	*Listeria monocytogenes*	Radiofrequency	Adherence of microbes on the surface reduced to 90% when compared to unmodified stainless-steel surface	Wang et al. (2003)
Polyethylene and stainless-steel surfaces	*E. coli*	Radiofrequency	Reduction of 7 log using water vapor as a source with an operating temperature did not rise beyond 44° C	Sen and Mutlu (2013)

has greater potential than direct application of plasma on microbial decontamination since it has ease of handling with specified dosage, storage capacity, and offsite generation. It has the same effect as that of ozone-treated water; here, ozone molecules are generated due to bubbling. The reactive species present in the water produced after the treatment act as an antimicrobial agent, which is utilized as an effective disinfectant or cleaning agent in equipment cleaning on the food contact surface. Some of the effects of PAW on microbial contamination are discussed in Table 11.2.

Handorf et al. (2020) studied the impact of milk spoiling microbes *Pseudomonas fluorescens* using microwave-induced plasma-activated water on the biofilm surface. Plasma-treated air and water are considered promising approaches for the application of industrial food decontamination. Plasma-activated water was used as a treating agent. At 300 s, pretreatment and 5 min of post-treatment activity on *Pseudomonas fluorescens* biofilm showed a 6 log reduction because of the antimicrobial effect of reactive species. Microbial cells became inactive up to 92% with a combination of 900 s of pretreatment and 5 min of post-treatment with plasma-activated water. Xiang et al. (2018) explored the effect of PAW using an atmospheric pressure plasma jet for the deactivation of *Pseudomonas deceptionensis* CM2. Exposing *P. deceptionensis* for 10 min to PAW, a 5 log reduction was achieved. PAW can be an alternative sanitizer for food processing equipment as the produced reactive species such as H_2O_2, nitrate, and nitrite anions have a significant effect on microbial inactivation.

TABLE 11.2   EFFECT OF PLASMA-ACTIVATED WATER IN CLEANING AND SANITATION

Medium of Plasma	Method of Plasma Generation	Target Microbes	Effects	References
Activated water	Microwave driven	*Pseudomonas fluorescens*	92% microbial inactivity at 900 s pretreatment and 5 min of post-treatment at 1 slm, forward power of 80 W and reverse power of 20 W	Handorf et al. (2020)
Activated water	Atmospheric pressure plasma jet	*Pseudomonas deceptionensis* CM2 strain	5 log reduction within 10 min of treatment time	Xiang et al. (2018)
Activated water and mild heat (40 to 50° C)	–	*Saccharomyces cerevisiae*	4.4 log CFU/ml reduction in combination effect	Zhang et al. (2020)
Activated water and mild heat (40, 50, 60° C)	APPJ system based on gliding arc discharge	*E. coli* O157:H7	Combination of 60 s of plasma treatment and mild heat at 60° C for 4 min achieved 8.28 log CFU/mL	Xiang et al. (2019)
Activated water	Cylindrical dielectric barrier discharge and jet	*E. coli* and *S. aureus*	Maximum reduction of *E. coli* at 7.14 ± 0.14 and of *S. aureus* at 3.10 ± 0.26 for 6.5 min exposure time at 3 slm of argon gas using jet plasma	Royintarat et al. (2019)
Activated water	Dielectric barrier discharge with electrodes	*Staphylococcus aureus*	Bovine serum albumin influenced the log reduction	Zhang et al. (2016)

Zhang et al. (2020) show the combined effect of PAW along with mild heat treatment (40–50° C) and individual treatment effect on the *Saccharomyces cerevisiae*. A combination of mild heat and PAW treated for 6 min garnered a 4.40 log CFU/mL reduction of *S. cerevisiae*. Alternatively, PAW treatment alone exhibits only a 0.27 log CFU/mL reduction for the same treatment time, whereas mild heat (50° C) treatment exhibited a reduction of 1.92 log CFU/mL. Xiang et al. (2019) also studied the effect of plasma-activated water and mild heat treatment (40, 50, and 60° C) on *E. coli* O157:H. A combination of 60 s of plasma treatment and mild heat at 60° C for 4 min shows a reduction of 8.28 log CFU/mL using an atmospheric pressure plasma jet (APPJ) system based on a gliding arc discharge.

Royintarat et al. (2019) examined the antibacterial effect of reactive species produced by plasma-activated water (PAW) using DC positive flyback transformer (FBT) jet plasma and cylindrical dielectric barrier discharge (C-DBD) with a neon transformer. In both the treatments, argon gas at a varied rate was used as a treatment gas for producing reactive species. Using jet plasma (FBT), the maximum reduction of *E. coli* was 7.14 ± 0.14 log and of *S. aureus* was 3.10 ± 0.26 log for 6.5 min exposure time at 3 slm. Using discharge plasma (CDBD), the maximum reduction of *E. coli* was 0.45 ± 0.07 log and of *S. aureus* was 2.45 ± 0.23 log for 11.5 min exposure time at 4 slm. Zhang et al. (2016) inferred that the presence of bovine serum albumin (BSA) exhibits a 2.1 to 5.5 log reduction of *Staphylococcus aureus* using plasma-activated water and in the absence of BSA, it shows a maximum inactivation of 7 log. In any case, the sterilization effect of PAW largely depends on the PAW treatment time of microbes and the time required for plasma generation.

11.6 SCOPE AND FUTURE TRENDS OF COLD PLASMA TECHNIQUES IN DAIRY PROCESSING EQUIPMENT

Nowadays, Zhu et al. (2020) reported that biofilm infection had become a great challenge in food safety. In dairy industries, the critical parameters with regard to the impact of biofilm formation are the presence of pathogenic microbes, the corrosion of metal surfaces, and changes in the organoleptic properties of food due to lipases or proteases. There are numerous manufacturing machinery and processes available in the milk processing line, such as raw milk tanks, pipelines, butter centrifuges, cheese tanks, pasteurizers, and packing tools, etc. These systems act as an active source of the biofilms of different microbial colony-forming species on the adherent surface of the machinery at different temperatures (Galié et al., 2018). Thomas and Sathian (2004) mentioned different types of cleaning agents, sanitizers, and disinfectants are being used for the CIP system. It can be stored and reused for cleaning, but repeated usage may cause the deposition of thermoduric organisms on the equipment's surface. This leads to the contamination of pasteurized milk. Good quality produce can only be available with proper supervision and adequate cleaning procedures in the food industry. An alternative approach to CIP cleaning is represented in Figure 11.2.

To overcome these challenges, the cold plasma technique could be an applicable futuristic technology with minimal or residual-free decontamination that acts as a disinfecting agent in the form of plasma-activated water on the food contact surface in the processing line. Usually, no industry would be prepared to modify existing lines and use plasma alone in the first few years as a decontamination procedure. The industrial plasma source must also be built so as to be applied in combination with existing lines and decontamination methods. Shen et al. (2016) and Schnabel et al. (2016) suggested that plasma-activated water is an indirect method of cold plasma

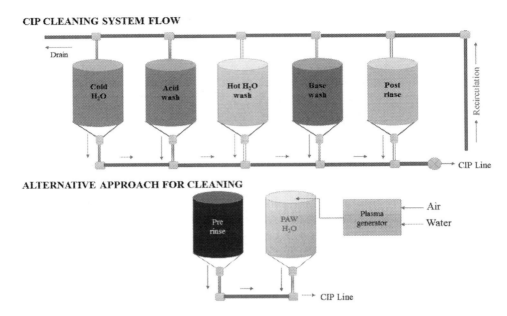

Figure 11.2 Pictorial representation of alternative approach to CIP cleaning.

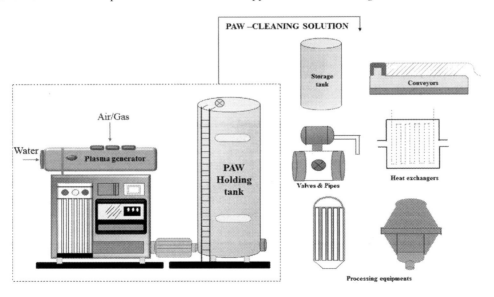

Figure 11.3 Conceptual design for cleaning of dairy equipment using plasma-activated water.

treatment that acts as a liquid decontamination agent for cleaning food processing equipment by the application of washing, spraying, or mists/gases. A conceptual design of a plasma system for the cleaning of dairy equipment using plasma-activated water is presented in Figure 11.3.

An attempt was made at the University of Liverpool, London, to scale up plasma treatment from lab level to a prototype for processing equipment cleaning in the food industry using a surface barrier discharge mechanism, which was found to be effective against pathogenic microorganisms (Walsh, 2017). For the scaling-up of lab-based research on cold plasma toward industry application as a novel anti-biofilm treatment, certain factors need to be considered like the type of gas involved, the method of plasma discharge, the nature of the treatment surface, operating conditions, target microorganisms, and the thickness of biofilms developed and food matrices, etc. Although the operating costs of the atmospheric pressure plasma jet cannot be compared to the other low-cost methods for decontaminating food contact surfaces, the potential for using this technology is still attractive and it does not release harmful byproducts. Some of the challenges that need to be addressed by the industry when implementing cold plasma technologies are establishing appropriate atmospheric plasma resources with proper monitoring systems, and the economic integration of plasma regulatory conditions with different research groups to successfully evaluate the efficiency and threat of any degradation by adjusting process parameters (power).

11.7 CONCLUSION

Cold plasma is discovered to be an excellent alternative novel non-thermal technique that tends to have increased attention and promising effects on cleaning, decontamination, and sterilization. Since cold plasma has a potential microbicidal effect on spoilage-causing biofilms, this makes it suitable to act as a disinfecting agent in dairy processing equipment; in short, as suggested in this chapter, plasma-treated water can be an attractive method for creating biologically active solutions that cannot easily be replicated with chemistry alone. Therefore, this methodology needs to be explored and studied as a green alternative or complementary technique to traditional methods of chemical processing. It can be further optimized for its functionality and specificity by deepening an understanding of basic chemical processes that govern plasma reactive chemistry. Future challenges are to introduce technological integration and upgrade to industrial processes with self-sufficient supply, on-demand, and simultaneous resource conservation. Regardless of the type of plasma-generating system, plasma techniques using indirect plasma (PAW) can be competitive with other surface sterilization methods and disinfection. Thus, this chapter highlighted the different approaches toward equipment disinfection and sterilization using plasma-activated water on dairy processing equipment, and also the disinfection effect of cold plasma on various food contact surfaces.

REFERENCES

Bayoumi, M. A., R. M. Kamal, S. F. Abd El Aal, and E. I. Awad. 2012. "Assessment of a Regulatory Sanitization Process in Egyptian Dairy Plants in Regard to the Adherence of Some Food-borne Pathogens and Their Biofilms." *International Journal of Food Microbiology* 158(3): 225–31. https://doi.org/10.1016/j.ijfoodmicro.2012.07.021.

Chen, Chen, Chenghui Liu, Aili Jiang, Qingxin Guan, Xiaoyuan Sun, Sisi Liu, Kexin Hao, and Wenzhong Hu. 2019. "The Effects of Cold Plasma-Activated Water Treatment on the Microbial Growth and Antioxidant Properties of Fresh-Cut Pears." *Food and Bioprocess Technology* 12(11): 1842–51. https://doi.org/10.1007/s11947-019-02331-w.

Cullen, K. L., E. Irvin, A. Collie, F. Clay, U. Gensby, P. A. Jennings, S. Hogg-Johnson, V. Kristman, M. Laberge, D. McKenzie, S. Newnam, A. Palagyi, R. Ruseckaite, D. M. Sheppard, S. Shourie, I. Steenstra, D. Van Eerd, and B. C. Amick. 2018. "Effectiveness of Workplace Interventions in Return-To-Work for Musculoskeletal, Pain-Related and Mental Health Conditions: An Update of the Evidence and Messages for Practitioners." *Journal of Occupational Rehabilitation* 28(1): 1–15. https://doi.org/10.1002/ppap.201700085.

Dasan, B. G., B. Onal-Ulusoy, J. Pawlat, J. Diatczyk, Y. Sen, and M. Mutlu. 2017. "A New and Simple Approach for Decontamination of Food Contact Surfaces with Gliding Arc Discharge Atmospheric Non-Thermal Plasma." *Food and Bioprocess Technology* 10(4): 650–61. https://doi.org/10.1007/s11947-016-1847-2.

Feizollahi, E., N. N. Misra, and M. S. Roopesh. 2021. "Factors Influencing the Antimicrobial Efficacy of Dielectric Barrier Discharge (DBD) Atmospheric Cold Plasma (ACP) in Food Processing Applications." *Critical Reviews in Food Science and Nutrition* 61(4): 666–89. https://doi.org/10.1080/10408398.2020.1743967.

Gabriel, Alonzo A., Maria Chelsea Clarisse F. Ugay, Maria Auxilla T. Siringan, Leo Mendel D. Rosario, Roy B. Tumlos, and Henry J. Ramos. 2016. "Atmospheric Pressure Plasma Jet Inactivation of Pseudomonas Aeruginosa Biofilms on Stainless Steel Surfaces." *Innovative Food Science and Emerging Technologies* 36: 311–19. https://doi.org/10.1016/j.ifset.2016.07.015.

Galié, Serena, Coral García-Gutiérrez, Elisa M. Miguélez, Claudio J. Villar, and Felipe Lombó. 2018. "Biofilms in the Food Industry: Health Aspects and Control Methods." *Frontiers in Microbiology* 9(May): 1–18. https://doi.org/10.3389/fmicb.2018.00898.

Greene, Annel K., Brian K. Few, and Joao C. Serafini. 1993. "A Comparison of Ozonation and Chlorination for the Disinfection of Stainless Steel Surfaces1." *Journal of Dairy Science* 76(11): 3617–20. https://doi.org/10.3168/jds.S0022-0302(93)77702-4.

Handorf, Oliver, Viktoria Isabella Pauker, Uta Schnabel, Thomas Weihe, Eric Freund, Sander Bekeschus, Katharina Riedel, and Jörg Ehlbeck. 2020. "Characterization of Antimicrobial Effects of Plasma-Treated Water (PTW) Produced by Microwave-Induced Plasma (MidiPLexc) on Pseudomonas fluorescens Biofilms." *Applied Sciences (Switzerland)* 10(9). https://doi.org/10.3390/app10093118.

Iuchi, Katsuya, Yukina Morisada, Yuri Yoshino, Takahiro Himuro, Yoji Saito, Tomoyuki Murakami, and Hisashi Hisatomi. 2018. "Cold Atmospheric-Pressure Nitrogen Plasma Induces the Production of Reactive Nitrogen Species and Cell Death by Increasing Intracellular Calcium in HEK293T Cells." *Archives of Biochemistry and Biophysics* 654: 136–45.

Jiménez-Pichardo, R., C. Regalado, E. Castaño-Tostado, Y. Meas-Vong, J. Santos-Cruz, and B. E. García-Almendárez. 2016. "Evaluation of Electrolyzed Water as Cleaning and Disinfection Agent on Stainless Steel as a Model Surface in the Dairy Industry." *Food Control* 60: 320–8. https://doi.org/10.1016/j.foodcont.2015.08.011.

Keidar, M., R. Walk, A. Shashurin, P. Srinivasan, A. Sandler, S. Dasgupta, R. Ravi, R. Guerrero-Preston, and B. Trink. 2011. "Cold Plasma Selectivity and the Possibility of a Paradigm Shift in Cancer Therapy." *British Journal of Cancer* 105(9): 1295–301. https://doi.org/10.1038/bjc.2011.386.

Liao, Xinyu, Yuan Su, Donghong Liu, Shiguo Chen, Yaqin Hu, Xingqian Ye, Jun Wang, and Tian Ding. 2018. "Application of Atmospheric Cold Plasma-Activated Water (PAW) Ice for Preservation of Shrimps (Metapenaeus ensis)." *Food Control* 94: 307–14. https://doi.org/10.1016/j.foodcont.2018.07.026.

Los, Agata, Dana Ziuzina, Daniela Boehm, Lu Han, Denis O'Sullivan, Liam O'Neill, and Paula Bourke. 2019. "Efficacy of Cold Plasma for Direct Deposition of Antibiotics as a Novel Approach for Localized Delivery and Retention of Effect." *Frontiers in Cellular and Infection Microbiology* 9: 428.

Mahendran, R., and K. Alagusundaram. 2015. "Uniform Discharge Characteristics of Non-Thermal Plasma for Superficial Decontamination of Bread Slices." *International Journal of Agricultural Science and Research* 5(2): 209–12.

Memisi, Nurgin, Slavica Veskovic Moracanin, Milan Milijasevic, Jelena Babic, and Dragutin Djukic. 2015. "CIP Cleaning Processes in the Dairy Industry." *Procedia Food Science* 5: 184–86. https://doi.org/10.1016/j.profoo.2015.09.052.

Mir, S. A., M. W. Siddiqui, B. N. Dar, M. A. Shah, M. H. Wani, S. Roohinejad, G. A. Annor, K. Mallikarjunan, C. F. Chin, and A. Ali. 2020. "Promising Applications of Cold Plasma for Microbial Safety, Chemical Decontamination and Quality Enhancement in Fruits." *Journal of Applied Microbiology*. https://doi.org/10.1111/jam.14541.

Muranyi, P., J. Wunderlich, and M. Heise. 2008. "Influence of Relative Gas Humidity on the Inactivation Efficiency of a Low Temperature Gas Plasma." *Journal of Applied Microbiology* 104(6): 1659–66. https://doi.org/10.1111/j.1365-2672.2007.03691.x.

Niemira, Brendan A., Glenn Boyd, and Joseph Sites. 2014. "Cold Plasma Rapid Decontamination of Food Contact Surfaces Contaminated with Salmonella Biofilms." https://doi.org/10.1111/1750-3841.12379.

Paatre Shashikanthalu, Sharanyakanth, Lokeswari Ramireddy, and Mahendran Radhakrishnan. 2020. "Stimulation of the Germination and Seedling Growth of Cuminum Cyminum L. Seeds by Cold Plasma." *Journal of Applied Research on Medicinal and Aromatic Plants* 18. https://doi.org/10.1016/j.jarmap.2020.100259. 100259.

Pankaj, S. K., C. Bueno-Ferrer, N. N. Misra, V. Milosavljević, C. P. O'Donnell, P. Bourke, K. M. Keener, and P. J. Cullen. 2014. "Applications of Cold Plasma Technology in Food Packaging." *Trends in Food Science and Technology* 35(1): 5–17. https://doi.org/10.1016/j.tifs.2013.10.009.

Park, Shin Young, and Sang-Do Ha. 2018. "Assessment of Cold Oxygen Plasma Technology for the Inactivation of Major Foodborne Viruses on Stainless Steel." *Journal of Food Engineering* 223: 42–5.

Pedrow, Patrick, Zi Hua, Shuzheng Xie, and Mei-Jun Zhu. 2020. "Engineering Principles of Cold Plasma." In: *Advances in Cold Plasma Applications for Food Safety and Preservation*, 3–48. Academic Press, 2020, London, United Kingdom.

Ratish Ramanan, K., R. Sarumathi, and R. Mahendran. 2018. "Influence of Cold Plasma on Mortality Rate of Different Life Stages of Tribolium castaneum on Refined Wheat Flour." *Journal of Stored Products Research* 77: 126–34. https://doi.org/10.1016/j.jspr.2018.04.006.

Royintarat, Tanitta, Phisit Seesuriyachan, Dheerawan Boonyawan, Eun Ha Choi, and Wassanai Wattanutchariya. 2019. "Mechanism and Optimization of Non-Thermal Plasma-Activated Water for Bacterial Inactivation by Underwater Plasma Jet and Delivery of Reactive Species Underwater by Cylindrical DBD Plasma." *Current Applied Physics* 19(9): 1006–14. https://doi.org/10.1016/j.cap.2019.05.020.

Sainz-García, Ana, Ana González-Marcos, Rodolfo Múgica-Vidal, I. Muro-Fraguas, R. Escribano-Viana, L. González-Arenzana, I. López-Alfaro, F. Alba-Elías, and E. Sainz-García. 2021. "Application of atmospheric pressure cold plasma to sanitize oak wine barrels" *LWT - Food Science and Technology*. https://doi.org/10.1016/j.lwt.2020.110509. 110509.

Sakudo, Akikazu, Tatsuya Misawa, and Yoshihito Yagyu. 2019. "Equipment Design for Cold Plasma Disinfection of Food Products." In: *Advances in Cold Plasma Applications for Food Safety and Preservation*, 289–307. Elsevier Inc. https://doi.org/10.1016/B978-0-12-814921-8.00010-4.

Sarmiento, Paula Marielisa, L. Luis Marcelo, and Jorge Isaac Fajardo. 2012. "Efficiency of the Low-Pressure Cold Plasma in the Cleaning of Steel Sheet for Subsequent Covering." https://doi.org/10.1109/Andescon.2012.35.

Sarmiento, Paula Marielisa, Luis Marcelo Lopez, Andres Paul Sarmiento, and Jorge Isaac Fajardo. 2012. "Efficiency of the Low-Pressure Cold Plasma in the Cleaning of Steel Sheet for Subsequent Covering." *Proceedings of the 6th Andean Region International Conference*, Andescon 2012: 115–18(November). https://doi.org/10.1109/Andescon.2012.35.

Schnabel, Uta, Rijana Niquet, Christian Schmidt, Jörg Stachowiak, Oliver Schlüter, Mathias Andrasch, and Jörg Ehlbeck. 2016. "Antimicrobial Efficiency of Non-Thermal Atmospheric Pressure Plasma Processed Water (PPW) against Agricultural Relevant Bacteria Suspensions." *International Journal of Environmental & Agriculture Research* 2: 212.

Scholtz, Vladimir, Jarmila Pazlarova, Hana Souskova, Josef Khun, and Jaroslav Julak. 2015. "Nonthermal Plasma: A Tool for Decontamination and Disinfection." *Biotechnology Advances* 33(6): 1108–19.

Şen, Yasin, Ufuk Bağcı, H. A. Güleç, and M. Mutlu. 2012. "Modification of Food-Contacting Surfaces by Plasma Polymerization Technique: Reducing the Biofouling of Microorganisms on Stainless Steel Surface.". *Food and Bioprocess Technology* 5(1): 166–75.

Şen, Yasin, and Mehmet Mutlu. 2013. "Sterilization of Food Contacting Surfaces via Non-Thermal Plasma Treatment: A Model Study with Escherichia coli-Contaminated Stainless Steel and Polyethylene Surfaces." *Food and Bioprocess Technology* 6(12): 3295–304. https://doi.org/10.1007/s11947-012-1007-2.

Shen, Jin, Ying Tian, Yinglong Li, Ruonan Ma, Qian Zhang, Jue Zhang, and Jing Fang. 2016. "Bactericidal Effects against S. aureus and Physicochemical Properties of Plasma Activated Water Stored at Different Temperatures." *Scientific Reports* 6(1): 1–10.

Sousa, J. S., K. Niemi, L. J. Cox, Q. Th. Algwari, T. Gans, and D. O'Connell. 2011. "Cold Atmospheric Pressure Plasma Jets as Sources of Singlet Delta Oxygen for Biomedical Applications." *Journal of Applied Physics* 109(12). https://doi.org/10.1063/1.3601347.

Thirumdas, Rohit, Deepak Kadam, and U. S. Annapure. 2017. "Cold Plasma: An Alternative Technology for the Starch Modification." *Food Biophysics* 12(1): 129–39. https://doi.org/10.1007/s11483-017-9468-5.

Thomas, Amitha, and C. T. Sathian. 2004 "Cleaning-in-Place (CIP) System in Dairy Plant-Review." *IOSR Journal of Environmental Science, Toxicology and Food Technology* 8(6): 41–44. https://doi.org/10.9790/2402-08634144.

Turner, M. 2016. "Physics of Cold Plasma." In: *Cold Plasma in Food and Agriculture*, 17–51, Academic Press, London, United Kingdom.

Varga, László, and Jeno Szigeti. 2016. "Use of Ozone in the Dairy Industry: A Review." *International Journal of Dairy Technology* 69(2): 157–68. https://doi.org/10.1111/1471-0307.12302.

Walsh, J. 2017. "IAFPs: European Symposium on Food Safety." Brussels, Belgium.

Wang, Y., E. B. Somers, S. Manolache, F. S. Denes, and A. C. L. Wong. 2003. "Cold Plasma Synthesis of Poly(Ethylene Glycol)-Like Layers on Stainless-Steel Surfaces to Reduce Attachment and Biofilm Formation by Listeria monocytogenes." *Journal of Food Science* 68(9): 2772–79. https://doi.org/10.1111/j.1365-2621.2003.tb05803.x.

Witte, Anna Kristina, Martin Bobal, Roland David, Beat Blättler, Dagmar Schoder, and Peter Rossmanith. 2017. "Investigation of the Potential of Dry Ice Blasting for Cleaning and Disinfection in the Food Production Environment." *LWT* 75: 735–41. https://doi.org/10.1016/j.lwt.2016.10.024.

Xiang, Qisen, Chaodi Kang, Liyuan Niu, Dianbo Zhao, Ke Li, and Yanhong Bai. 2018. "Antibacterial Activity and a Membrane Damage Mechanism of Plasma-Activated Water against Pseudomonas Deceptionensis CM2." *LWT* 96: 395–401. https://doi.org/10.1016/j.lwt.2018.05.059.

Xiang, Qisen, Wenjie Wang, Dianbo Zhao, Liyuan Niu, Ke Li, and Yanhong Bai. 2019. "Synergistic Inactivation of Escherichia coli O157:H7 by Plasma-Activated Water and Mild Heat." *Food Control* 106. https://doi.org/10.1016/j.foodcont.2019.106741. 106741.

Zhang, R., Y. Ma, D. I. Wu, L. Fan, Y. Bai, and Q. Xiang. 2020. "Synergistic Inactivation Mechanism of Combined Plasma-Activated Water and Mild Heat against Saccharomyces Cerevisiae." *Journal of Food Protection* 83(8): 1307–14.

Zhang, Qian, Ruonan Ma, Ying Tian, Bo Su, Kaile Wang, S. Yu, Jue Zhang, and Jing Fang. 2016. "Sterilization Efficiency of a Novel Electrochemical Disinfectant against Staphylococcus Aureus." *Environmental Science and Technology* 50(6): 3184–92. https://doi.org/10.1021/acs.est.5b05108.

Zhu, Yulin, Changzhu Li, Haiying Cui, and Lin Lin. 2020. "Feasibility of Cold Plasma for the Control of Biofilms in Food Industry." *Trends in Food Science and Technology*. https://doi.org/10.1016/j.tifs.2020.03.001.

Zhukorskyi, O., and Y. Kryvokhyzha. 2016. "Ecological Risks of Using Chemical Sanitizing Agents for Milking Machines and Milk Containers." *Agricultural Science and Practice* 3 (3): 12–16. https://doi.org/10.15407/agrisp3.03.012.

Chapter 12

Safety, Regulatory Aspects and Environmental Impacts of Using Non-Thermal Processing Techniques for Dairy Industries

Khalid A. Alsaleem, Ahmed R. A. Hammam and Nancy Awasti

CONTENTS

12.1 Introduction	157
12.2 Historical Overview	160
12.3 Commercial and Regulatory Requirements	162
12.4 Environmental Impact	165
12.5 Energy Savings	166
12.6 Reduced Gas and Water-Saving	167
12.7 Conclusions	168
References	168

12.1 INTRODUCTION

Consumers require healthy, safe, fresh-like, and chemical-free foods with a high nutritional value. To meet consumer requirements, many technologies have been utilized to preserve foods. Non-thermal technologies are one of the techniques that have been widely used recently. The non-thermal process is a technology that aims to inhibit microorganisms and enzymes while increasing the storage stability or shelf-life of foods without using heat treatments. As a result, compared to existing thermal processes, the impact of non-thermal technology on the characteristics of food products (e. g. flavor, color, nutritional values) is low. In contrast, the products meet the consumer requirement with higher standards of food safety. Additionally, non-thermal technology requires shorter times than thermal technologies, such as pasteurization, evaporation, and drying (Zhang et al., 2019).

Several non-thermal technologies have been used, including supercritical fluid technology (SFE), ultrasound (US), irradiation (IR), membrane filtration, cold plasma (CP), hurdle technology (HUT), pulsed electric fields (PEF), pulsed UV light (PL), hydrodynamic cavitation processing (HCP), high-pressure processing (HPP), and ozone (Table 12.1). Those techniques have shown

TABLE 12.1   THE PRESERVATION OF NON-THERMAL TECHNOLOGIES IN FOOD PROCESSING

Process[1]	Description	Applications	References
SFE	Using high-pressure (5–30 MPa) carbon dioxide at low temperature (cold pasteurization)	Liquid foods Extract functional components	(Garcia-Gonzalez et al., 2007; Martín-Belloso et al., 2014)
Membrane filtration	Using filtration systems such as microfiltration membranes with 0.1 and 10 µm pore size to fractionate liquid components	Liquid foods: milk, beer, wine, and fruit juice	(Moraru & Schrader, 2008)
HUT	Combination of techniques applied to food, and process intrinsic and extrinsic factors	Fermented foods Cured meats Fruit preserves or jams demonstrate	(Leistner & Gould, 2002)
PL	Emits a range of 1 to 20 flashes per second at a range of energy density from 0.01 to 50 J/cm^2	Surface of food Equipment Food packaging materials	(Cheigh et al., 2012; Gayán et al., 2014)
US	Frequency > 100 kHz at intensity < 1 W cm^{-2} or frequency ranges from 18 to 100 kHz at intensity > 1 W/cm^2	Food emulsification Sterilization Extraction Freezing fresh foodstuffs	(Aadil et al., 2015; Chemat et al., 2011)
IR	The source of energy gamma rays is ^{60}Co, and sterilization doses range from 1 to 50 kGy	Prepackaged foods Bulk foodstuffs	(Farkas, 2006; Feliciano et al., 2014)
CP	The barrier glow discharge between two parallel electrodes	The surface of raw produce Packaging materials	(Fernandes et al., 2012; Tani et al., 2012)
PEF	Electric field intensity ranges from 20 to 80 kV/cm with few pulses with < 1 s time	Liquid foods: fruit juice and milk	(Sharma et al., 2014; Toepfl et al., 2006)
HCP	Works using pressure (2 to 15 bar) for 5 to 25 min to inactivate microorganisms and enzymes	Liquid foods: fruit juice and milk	(Arya et al., 2020; Katariya et al., 2020; Li et al., 2018; Salve et al., 2019)
HPP	Pressure ranges from 200 to 400 MPa with temperature < 50° C	Prepackaged foods Bulk foodstuffs Fruit juices or jams	(Caminiti et al., 2011; Rendueles et al., 2011; Zhao et al., 2019)
Ozone	Allotropic form of oxygen and has different effects depending on the concentration (ppm) and exposed time	Antimicrobial agents Fruit juices	(Cullen et al., 2010)

Notes

[1] Process: SFE = supercritical fluid technology; HUT = hurdle technology; PL = pulsed UV light; US = ultrasound; IR = irradiation; CP = cold plasma; PEF = pulsed electric fields; HCP = hydrodynamic cavitation processing; HPP = high-pressure processing.

practical impacts on preserving foods from spoilage and pathogenic microorganisms and have a wide range of food applications (liquid foods, fermented foods, meat, prepackaged foods, etc.) and non-food applications (food packages, surfaces, cleaning agents, etc.).

The SFE method is well known as cold pasteurization and is typically used in liquid foods using carbon dioxide and pressures that range from 5 to 30 MPa (Garcia-Gonzalez et al., 2007; Martín-Belloso et al., 2014). There is also membrane filtration, such as microfiltration, ultrafiltration, and nanofiltration membranes, used in liquid foods, especially in milk and fruit juices (Moraru & Schrader, 2008). Each membrane type has a specific pore size based on the size of components that need to be fractionated. Another non-thermal process is HUT, which applies a combination of fermented foods, cured meats, and fruit preserves (Leistner & Gould, 2002). Some non-thermal technologies, such as the PL, are used to sterilize the surface of foods, equipment, as well as food packaging material using 0.01 to 50 J/cm² flashes (Cheigh et al., 2012; Gayán et al., 2014), while CP is applicable in packaging materials using two parallel electrodes (Fernandes et al., 2012; Tani et al., 2012). On the other hand, PEF uses electric field intensity (range from 20 to 80 kV/cm) with pulses in less than one second in liquid foods (Sharma et al., 2014; Toepfl et al., 2006). For US, frequency (18 to 100 kHz) is used at the intensity of 1 W/cm² or higher, and it is utilized for food emulsification, sterilization, extraction, as well as freezing fresh foodstuffs (Aadil et al., 2015; Chemat et al., 2011). Figure 12.1 shows a schematic diagram of the ultrasonic measurement system on a lab scale. IR is also applied in prepackaged foods and bulk foodstuffs as a non-thermal process using gamma rays (Farkas, 2006; Feliciano et al., 2014). HCP is also considered a non-thermal process, which is widely utilized in liquids such as milk and juices by exposing the product to a range of pressures (3 to 15 bar) for a specific time (Arya et al., 2020; Katariya et al., 2020; Li et al., 2018; Salve et al., 2019). Additionally, ozone (an allotropic form of oxygen) is applied as an antimicrobial agent, as well as in fruit juices (Cullen et al., 2010). HPP is the most common non-thermal process among those non-thermal technologies and is utilized in many food applications (jams, fruit juices, etc.) (Caminiti et al., 2011; Rendueles et al., 2011; Zhao et al., 2019).

Some non-thermal techniques are not efficient for preserving foods from pathogenic or spoilage microorganisms and enzymes. As a result, one technique is merged or combined with another process (such as HPP with heat, low pH, modified atmospheres, carbon dioxide, and

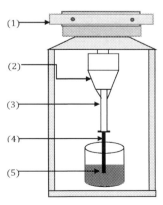

Figure 12.1 Schematic diagram of the ultrasonic measurement system: (1) generator, (2) transducer, (3) booster/amplifier, (4) sonotrode/horn, (5) milk sample.

antimicrobials; US with low pH, low a_w, and heat; PEF with heat, low pH, antimicrobials, and HPP; IR with low temperature and modified atmospheres, heat, and HPP) to eliminate those microorganisms and keep the foods safe, which increases the shelf-life (Raso & Barbosa-Cánovas, 2003).

12.2 HISTORICAL OVERVIEW

The milestones of non-thermal techniques in the food and dairy industry are shown in Table 12.2. As early as 1903, UV light had been examined on bacteria (Rentschler et al., 1941), while in 1932, it was found that the total bacterial count and *E. coli* count were reduced in milk and other liquids when they were exposed to an 8,900-cycle frequency of audible sound waves (Chambers & Gaines, 1932). After nearly three years, Sluder used infrared lamps in a vacuum-belt drier to dehydrate orange juice (Sluder et al., 1947). Thereafter in 1956, a non-thermal drying method was applied to *Ascophyllum nodosum* and the stipe of certain sublittoral seaweeds (brown marine algae) to reduce the water content to 50% (Reid & Jackson, 1956). In 1960, radiation was utilized for the sterilization of milk (Wertheim, 1960), while a new method was invented for the irradiative treatment of milk and other food products in 1976 (De Stoutz, 1976).

In 1990, Lammerding and Doyle studied different non-thermal conditions, including low pH and low temperature and traditional and novel antimicrobial compounds on the existence of *Listeria monocytogenes* in foods (Lammerding & Doyle, 1990). After that, another study examined the effect of PEF non-thermal process on the inactivation of *Saccharomyces cerevisiae* in apple juice in 1995 (B.-L. Qin et al., 1995). Another study used high-intensity PEF to inhibit microorganisms and enzymes with a small elevation in the temperature of food, as well as enhancing the economical cost with practical usage of energy (B. Qin et al., 1996). In 1998, Barnett mentioned different types of cavitation that could be utilized in the food industry (Barnett, 1998). It has been found that single non-thermal technology is not enough to increase food safety, so Raso and Barbosa-Cánovas recommended in 2003 to merge a non-thermal technology with other processes to preserve foods and increase shelf-life by eliminating microorganisms and enzymes (Raso & Barbosa-Cánovas, 2003).

Five years later, Unluturk used UV light radiation to decrease *E. coli* strain counts and found that this technology could be used to pre-treat liquid egg products by itself or combined with mild heat treatment. This can be used to avoid the adverse effects of pasteurization (Unluturk et al., 2008). In 2015, a CP technique was used to inhibit microorganisms and enzymes responsible for food browning during heat treatment to increase storage stability or shelf-life (Thirumdas et al., 2015). Recently, it has been reported that most non-thermal technologies are not enough to sterilize the products thoroughly, which can reduce food safety and shelf-life due to spoilage or pathogenic bacteria (Wu et al., 2020).

Non-thermal technologies are environmentally friendly due to their reduced energy and water consumption, plus emissions. Every non-thermal technology has strengths and weaknesses, which can affect its utilization in different applications. The HPP has a wide range of microorganisms and less effect on products, as well as being energy efficient. However, this technology is expensive and has less effect on spores (Zhao et al., 2019). It has been reported that the cost of HPP required to process orange juice is seven times higher than traditional thermal processing (1.5 ¢/l), while the CO_2 emission is approximately eight times high in HPP (773,000 kg) relative to the conventional thermal (90,000 kg) process (Sampedro et al., 2014). Although the production cost and environmental impacts of HPP are higher than conventional thermal technologies, its implementations in the food industry have widely increased (Sampedro et al., 2014). Cacace and others also found that HPP costs more as compared to the thermal process in treating orange

12.2 Historical Overview

TABLE 12.2 SELECTED SCIENTIFIC AND COMMERCIAL MILESTONES IN NON-THERMAL TECHNOLOGIES IN THE FOOD INDUSTRY

Year	Milestone	References
1903	UV light effects on bacteria were observed as early as 1903	(Rentschler et al., 1941)
1932	It is possible to reduce the counts of microorganisms such as *E. coli* in liquid by exposure to the sound of an audible frequency (8,900 cycles)	(Chambers & Gaines, 1932)
1947	Utilized infrared lamps in a vacuum-belt drier for the dehydration of orange juice	(Sluder et al., 1947)
1956	Used non-thermal drying method to reduce the water content to 50% in brown marine algae (*Ascophyllum nodosum* and the stipe of certain sublittoral seaweeds)	(Reid & Jackson, 1956)
1960	Used radiation sterilization of fluid food products, including milk	(Wertheim, 1960)
1976	A new method and apparatus was invented for the irradiative treatment of beverages, such as milk, beer, wine, and fruit juice	(De Stoutz, 1976)
1990	Studied several non-thermal processing conditions (low pH, low temp, traditional and novel antimicrobial compounds) on *Listeria monocytogenes* in foods	(Lammerding & Doyle, 1990)
1995	Used PEF as a non-thermal process to inhibit *Saccharomyces cerevisiae* in apple juice	(B.-L. Qin et al., 1995)
1996	Used high-intensity PEF to inhibit microorganisms and enzymes while improving the economic and efficient use of energy	(B. Qin et al., 1996)
1998	Mentioned several types of cavitation that can be applied in the food industry	(Barnett, 1998)
2003	The use of non-thermal processes in combination with other preservation technologies can increase the storage stability or shelf-life and food safety due to their effects on psychrophilic pathogens, spoilage micro-organisms, bacterial spores, and certain enzymes	(Raso & Barbosa-Cánovas, 2003)
2008	Found that UV light radiation reduced the *E. coli* strain counts and can be used as a pre-treatment technology or merged with mild heat treatment to lessen the undesirable effects of thermal pasteurization in liquid egg products	(Unluturk et al., 2008)
2015	CP reduces microorganisms, as well as inactivating enzymes that are responsible for food browning during heat treatment to extend the shelf-life	(Thirumdas et al., 2015)
2020	Most non-thermal technologies are unable to sterilize thoroughly, therefore some microbial spores or pathogens exist that can affect the food safety and shelf-life	(Wu et al., 2020)

juice; however, HPP is cheaper in processing sliced Parma ham. Nevertheless, they reported that HPP has less environmental impact relative to the thermal process and modified atmosphere packaging when orange juice or sliced Parma ham are processed (Cacace et al., 2020).

US technology has been described as a non-invasive and straightforward process with less pollution. However, its disinfecion ability is low (Zhao et al., 2019). It is applicable in different food applications, but its utilization in the industry is low due to the need for extra equipment for each application and the lack of collaboration between academia and industry (Compton et al., 2018; Knorr et al., 2011).

Another non-thermal process is PEF, which requires low energy and less time, but its impact on spores and viruses is low; additionally, it has a higher cost and more water usage (Zhao et al., 2019). Sampedro reported that the CO_2 emission is 700,000 kg in PEF as compared to 90,000 kg in the conventional thermal process (Sampedro et al., 2014). On the other hand, PL requires less cost and a suitable process for food surface, equipment, and packaging materials; but this technology negatively impacts the sensory properties of food (Zhao et al., 2019).

CP technology is also another non-thermal process that is used in the industry. It is useful and can eliminate a wide range of microorganisms at low temperatures. However, this process can affect the sensory quality and leave some residue on the food (Zhao et al., 2019). Another process is ozone technology that can inhibit several types of microorganisms with less cost. This technology is typically touching the food directly and does not leave any food residues. However, it does not apply to organic materials (Zhao et al., 2019).

12.3 COMMERCIAL AND REGULATORY REQUIREMENTS

With the increase in demand among the general public for minimal thermally processed food, non-thermal processes have raised continuous pressure on food regulators and equipment vendors. To maintain the product's high sensory and nutritional value, continuous efforts and studies are being made at the lab and pilot scale. Thermal treatments are common methods to increase the shelf-life of the food and food product, but these technologies are also associated with irreversible changes in the treated food's sensory and nutritive profile. For example, heating milk or food products at elevated temperatures of $\geq 100°$ C usually results in the loss of vitamins, flavor, aroma, etc. (Ajmal et al., 2018; Barrett et al., 2010; Kebede et al., 2014; Oupadissakoon et al., 2009). Therefore, nowadays, non-thermal treatment is gaining a lot of attention and demand as an alternative source of food preservation in the present market (Guerrero-Beltran & Barbosa-Canovas, 2005). Non-thermal processes or treatments such as irradiation, pulsed electric field, ozone, high-pressure processing, ultraviolet light, and pulsed light are a few technologies that inactivate or kill microorganisms in food (Pereira & Vicente, 2010). To produce a safe and better quality of food and food products, intense research is being performed to incorporate and seek a novel combination of already existing technologies. To prevent food safety hazards and retain the quality of foods and their products, several government and global standards have been implemented (Awasti et al., 2021) and thus should be followed during the integration of new technologies such as non-thermal treatments.

Although the above-mentioned non-thermal technologies have a great scope for producing safe and high-quality food and dairy products, the implementation of these processes has been held back from a complete industrial-scale capacity due to their existing limitations. One of the most crucial requirements for successfully implementing and commercializing new technologies is the grant of permission from the government and regulatory agencies (Han, 2009). The lack of scientific data on the commercialization of such technologies at the industrial and pilot-scale levels has been challenging. It thus requires further research and analysis underlining different variables with respect to the product and technology being tested. This section of the chapter mainly focuses on various aspects and requirements of the non-thermal process in terms of regulations and commercial acceptance of the technology. This section also provides a brief discussion of the different steps required to approve a new food processing technology.

One of the critical requirements for implementing a new food processing technology at the commercial level is permission from government or regulatory agencies. Since the process related to food legislation is very tedious and political, granting permission for regulatory issues to use a non-thermal process is also time-consuming and political. Simultaneously, different interested parties such as equipment vendors, consumer groups, the food industry, and regulatory organizations may have different opinions on the same problem, thus making acceptance of the technology more strenuous. In addition to this, there is not a sufficient amount of scientific data or research that can provide a solution to all these interested parties. Therefore it becomes very important to understand and identify the existing status of technologies at a commercial level and provide insight into development to academic and research societies.

Among most of the existing non-thermal processing techniques, pulsed light is not used much in the food industry at a smaller scale but has been successfully implemented to decontaminate bottle caps and bonding of CD/CVD/Blu-Ray. This technology was commissioned in the year 1995 by the Food and Drug Administration for its commercial-scale use in food for the decontamination of food packaging material. Thus, it can now be easily adapted and integrated by any food industry at a commercial level into their already existing processing lines.

On the other hand, numerous research studies have been performed on the pulsed electric field (PEF), but it has not become commercially successful and is mostly used at a pilot plant or experimental stage. The first PEF was implemented in the United States by a company named Genesis Juice Corporation (2005). They used this technology for non-thermal pasteurization of fruit juices with a capacity of 200 L/h of the processing unit (Clark, 2006). EF is mainly a physical process of preservation of food and falls under the regulation of the US Food and Drug Administration (FDA). Due to its low commercialization, this process has experienced difficulty against other existing thermal technologies. One of the major safety concerns of using PEF as a preservation method is the migration of electrode material from the treatment chamber to the food. Therefore, the FDA requires a petition and evaluation before approval of any non-thermal process such as PEF. At last, only after completion of the evaluation, a ruling is generated or issued which is then added to the codex of federal regulations, whereas for premarket approval of PEF as a food processing technology, in-depth scientific review with specific data is required for detailed characterization of the product and the process. The document submitted to the FDA should address some important details such as the type of product and process being analyzed, operational guidelines or a detailed step-by-step description on how to use the technology, and expected accomplishment after using the suggested technology on the product. Apart from the above-mentioned factors, the impact of the process implementation on the environment also plays a crucial role (Hansen, 1999). Thus, intense communication with the FDA should always be followed by the processors.

Presently, the basis for PEF technology exists and is strongly supported by numerous years of research and studies. Now it's time to conduct detailed and systematic research that addresses the questions of the FDA and various issues regarding the implementation of technology and food hazards. FDA regulations are mostly specific and are associated with United States regulations, but its existence in delivering safe food to customers and consumers goes far beyond the US's borders. In Japan and Germany, ozonized water is used on fishing boats routinely to wash and ice-pack fishes on board. On similar lines of Japanese approval for ozonized water, in 1996, the Australian government also approved the use of ozone on all types of food surfaces. In Norway, aquaculture farms and plants started to use ozone for preserving fish in ice packs made up of ionized water. Similarly, the Canadian Food Inspection Agency (CFIA) also approved the use of ozone as a cleaner that can be used directly on food surfaces. In continuation of this, the

FDA on June 26, 2001, authorized the use of a few antimicrobial agents in the US, such as ozone treatment for food and food contact surfaces. After FDA approval of using ozone as a treatment method, the Ministry of Agriculture of the US in December 2001 accepted its direct use in poultry, crustaceans, meat, and fish. The FDA also listed ozone under "generally recognized as safe" (GRAS) status.

In addition to ozone treatment, irradiation treatment is also widely accepted and approved by more than 40 countries for over 50 different types of food and food products. In 1997 (Merrill), food irradiation was assessed by the US Congress as a food additive. Based on national health authorities, every country has its own sets of regulations to use or approve any new technology. Some countries set up or create regulations depending on a case-by-case analysis, for example, based on research, analysis, and data, whereas many countries utilize the knowledge, recommendations, facts, and conclusions from various expert committees like WHO, FAO, and IAEA. Even so, the purpose and type of food being irradiated differ from country to country. The acceptance of the concept of "chemiclearance" by the FDA (US) was one of the most significant milestones in rulemaking for the approval of irradiation technology to produce irradiated food. According to labeling regulations and requirements, all irradiated food items except spices need to be labeled with the logo "Radura" and a phrase like "treated by irradiation for quarantine purpose" or "treated with radiation to control spoilage." In addition to this, the irradiated food produced by the manufacturers should also include the source of radiation by using labels such as "treated with ionizing radiations," "treated with gamma radiations," or treated with X-radiations." An additional statement can also be added by the manufacturer as a part of consumer knowledge like "This treatment does not include radioactivity" and must only be included if the statement is verifiable and does not misguide the consumer.

In the United States, the basis for promoting specific regulations by the FDA requiring processing and packaging under sanitary conditions is mostly driven by the Food, Drug, and Cosmetic Act (FD&C). HPP is a widely accepted technology and is used by manufacturers in producing packed fruits, seafood, vegetables, meat, and dairy products. HPP is a USDA-approved technology that uses high water pressure to defend against pathogens and spoilage, causing the death of microorganisms without altering the product's quality in terms of texture, taste, nutritional value, and appearance. Currently, products produced using HPP technology are usually distributed under cold or refrigerated conditions, similar to pasteurized products, and require mandatory GMP conditions and specific regulations based on the product being made such as the Pasteurized Milk Ordinance (PMO), Juice HACCP, or Seafood HACCP. During refrigerated storage and distribution, the potential temperature abuse must be minimized and controlled. A detailed review by Huang and others mentioned that currently, more than 300 sets of HPP equipment have been functioning worldwide, including Asia (12%), Europe (25%), and North America (54%) for mass production of HPP products (Huang et al., 2017). A recent study by Kumari and Farid reported improved microbiological quality of fruit pulp prepared using high-pressure processing (Kumari & Farid, 2020).

Overall, the mission of the FDA is to regulate and assure the protection of public health by keeping mandatory requirements for the commercial approval of any novel technology or process. However, the FDA is not involved in the approval process but usually deals with the components or ingredients used in the process. Before availing an approval from current regulators, a few technologies such as PEF and other emerging techniques such as cavitation (Awasti et al., 2019; Chaudhary, 2019) may also need a pre-market approval to be accepted as a non-thermal technology. Therefore, to comply with safety regulations laid out by the FDA, different types of food and food surfaces should be thoroughly studied at experimental and pilot-scale levels.

During process implementation or filing of a new process or technology, a series of steps should be followed such as establishing a continuous and thorough communication with the FDA during process development (Larkin & Spinak, 1996), describing and explaining the process in detail to the FDA, on-site evaluation of the product and process by inviting the FDA to the production facility, and proposing an outline of the process with detailed information on resistant microorganism viability and the least lethal zone of treatment in the system with regards to food safety and quality. Further, the FDA reviews the equipment design and filing technology for its process novelty. In total, the significant aspects should cover the issues in the area of microbiology, nutrition, and toxicology. Details submitted by the processor must address a few questions based on the quality and safety of food, such as a) whether there are any toxic compounds produced during the process due to changes in the product's texture, functionality, or color, b) how the technology being used ensures the sterility of unsterile products from microbes and toxins, and c) whether there are any quality or nutritional losses during or after the product processing. The processors should use approved material in making these technologies, and they should closely work with the equipment vendors to ensure the safety and quality of the product. Since 1997, EU regulations (CE 258/97) for novel food ingredients took hold on all novel applications related to precautionary principles for new products that were previously regulated by the national regulations. This now allows the legislations to establish an evaluation and certification or licensing for "novel foods" required by all new food processors and processes.

Overall, the regulations are the most crucial factors in determining the facilitation or limiting the use of filling technology, especially in the case of implementation of the non-thermal process. In-depth understanding and knowledge of operational guidelines, conditions during operation, compatibility of packaging material with that of the technology being produced, detailed evidence of functionality, safety, and the nutritional and sensory qualities of the products are a few important requirements that are needed to convince the regulatory officials and authorities. The requirements presented to the authorities should be able to prove and convince that these novel non-thermal processing technologies produce products that are safe and can be made available to the public via commercial production by using approved technology.

12.4 ENVIRONMENTAL IMPACT

The environmental impact is known as any modifications to the environment, whether negative or positive, related to human activities (Glasson & Therivel, 2019). Facility activities and products and services are examples of human activities that could result in acid rain, climate change, and ocean acidification. One of the causes of environmental impact is food production, and the energy used in production is the most critical factor that may imbalance the food balance in the future. Three axes should be considered to ensure the nutritional balance preservation with the increase in the earth's population: water, food, and energy. Reducing the adverse effects on the environment and the production cost are producers' and consumers' demands, leading to the search for alternative methods to the traditional production methods (Toepfl et al., 2006).

During the past three decades, the focus has been on the energy used in food production and environmental impact. Due to the increasing demand for food, the safety requirements in production, and cleaning cycles for manufacturing tools are required to process food that considers using a large amount of cold and hot water (Dalsgaard & Abbotts, 2003). This may increase water and energy consumption and wastewater. The traditional methods, known as thermal methods, used in dairy factories such as pasteurization, sterilization, and drying harm the environment.

The thermal treatments have low economic feasibility due to expensive cleaning, energy, human resources, and environmental impact (Sandu & Singh, 1991). Consumers are concerned about the quality of dairy products and want them to be treated as minimally as possible to obtain a longer shelf-life and be environmentally friendly (Mir et al., 2016).

12.5 ENERGY SAVINGS

The cost of fuel and electricity consumption for dairy factories in the USA is estimated at $1.5 billion annually (Masanet et al., 2014). Forty-eight percent of energy consumption went to electricity consumption, and 52% went to fuel consumption (US Department of Commerce, 1980). Eighty percent of fuels are used directly to produce heat and steam, 20% used for different processes such as HVAC systems (Masanet et al., 2014). Table 12.3 shows the estimated energy required for processing different dairy products using thermal treatments. In general, the sources of energy for these treatments are fossil fuel and electricity. The energy consumption in dairy factories is divided into 1) treatment processing, divided into pasteurization, sterilization, evaporation, and drying, and 2) electrical processing, including lighting, cooling, air systems, fans, freezing, and pumping (Dalsgaard & Abbotts, 2003). The source of fuel in dairy industries is natural gas. In 2006, the dairy industries used around 13% of food industries' natural gas consumption, which is about 80 million cubic feet (Administration, 2006). The operational cost and energy required to process food were not a concern before the oil crisis in 1970 (Batie & Healy, 1983). The energy resource prices are not constant, making it challenging for dairy plants to calculate each product's fixed value. Comparing energy prices in the United States from 1970 to 2009, fuel and electricity prices had increased by two- and 1.2-fold, respectively (Bureau, 2009).

The effectiveness of non-thermal treatment in saving energy is proven by reducing the amount of gas and electricity required to reach quality and safety specifications. A previous study shows the possibility of reducing the savings of 100% of gas and 18% of electricity to replace thermal processing techniques such as pasteurization and drying using non-thermal processing techniques such as pulsed electric fields in orange juice (Lung et al., 2006). The non-thermal treatments are carried out at a moderate temperature and do not require high costs for cooling the product after treatment. In contrast, the thermal treatments require fast cooling for products due to the high temperature used in the treatment and the quality of the product that could be affected (Cheftel, 1992). High hydrostatic pressure is one of the most popular non-thermal treatments used in dairy products, and it processes at room temperature (Cheftel, 1992). It generates heat

TABLE 12.3 ENERGY USED IN THERMAL TREATMENTS FOR DAIRY PRODUCTS

	Milk	Cottage Cheese and Yogurt	Cheese	Dry Whey	Evaporated Milk	Powdered Dry Milk	Ice Cream	Creamery Butter
Direct fuel	28	17		1,115		115	15	
Steam	117	43	356	3,881	312	205	92	359
Electricity for refrigeration		27	193	193			659	175
Other electricity		32	648	648	132	101	37	18

*Btu/lb of product.
Source: adapted from Brush et al. (2011).

for a product during processing due to the high pressure, which could be released by releasing the pressure (Shao et al., 2010). Using high hydrostatic pressure for sterilization could decrease energy requirements by 20% compared to the thermal treatments (Toepfl et al., 2006).

Using the non-thermal treatments alone or with thermal treatments has been shown to save energy. Traditional smokehouse cooking could be replaced using ohmic heating, saving more than 70% of energy (Vicente et al., 2006). Treating sweet potatoes with ohmic heating before freeze-drying could save more than 25% of energy (Lima et al., 2002; Masanet, 2008). One of the most critical difficulties facing traditional treatment is the fouling of heat exchangers that could increase pressure and decrease heat transfer efficiency (Toyoda et al., 1994). This could lead to a decrease in efficiency and an increase in the energy and cost by the additional workforce, chemicals, equipment, and environmental impact (Gillham et al., 2000).

12.6 REDUCED GAS AND WATER-SAVING

Recently, studies and efforts have been focused on developing the manufacture and limiting the use of non-renewable energy to reduce the waste of water, energy, and gas emission. The traditional technologies have environmental impacts due to their energy and water consumption and gas emissions. These technologies rely on temperature and time protocols that require a higher amount of fossil fuels and water. In 1997, the food industry ranked fifth of the US's largest industrial consumers, with 4.4% of energy consumption coming, after petroleum and coal products, chemicals, paper, and primary metals (Muller et al., 2007). In 1998, the energy used in the US food processing industry could be summarized as follows: 1) heating that used 29% of total energy input and could result in particulate matter, gases, and volatile organic compounds; 2) cooling and refrigeration that consumed 16% of total energy input and could produce halogenated volatile organic compounds; and 3) steam production that consumed 33% of the total energy input (Lung et al., 2006). The uses of steam in food factories center on cooking, evaporation, drying, and sterilization. In California food industries, 57% of fossil fuels are used to generate steam (Einstein et al., 2001). The traditional method used to produce steam in food factories is the use of boilers and freshwater. One of the disadvantages of the method is the number of impurities remaining in it and building up in the boiler, which periodically led to removing water from the boiler's bottom to discard the remainder of impurities (Pereira & Vicente, 2010). The removing water rate could be higher than 15% of seam productivity, which is greater than the recommended rate of 1–3% (Wang, 2008). The number of impurities remaining could decrease boiler efficiency by 20–30%, leading to wasting energy and more significant emissions of CO_2 (Pereira & Vicente, 2010). Flue gas, boiler blowdown water, incomplete combustion, fouling of heat transfer surfaces, and heat convection and radiation losses from the hot boiler surface are examples that could cause waste energy in the boiler (Wang, 2008). One of the dairy manufacturing process outputs is gas emissions. These gas emissions result from combustion, cooking and heating processes, and refrigeration systems (Pereira & Vicente, 2010). The gas emissions include carbon dioxide, sulfur dioxide, nitrogen oxides, and volatile organic compounds that could harm the environment. The amount of these emissions relies on the type of fuel used (Toogood & Key, 2000). Using non-thermal treatment instead of thermal treatment in dairy factories may reduce pollution and save the environment and energy due to reducing the need for cooling systems (Dalsgaard & Abbotts, 2003). Moreover, relying on electricity in dairy industries from renewable energy resources such as solar and wind electric and hydroelectric power could reduce the waste of water, energy, and gas emissions and save energy.

12.7 CONCLUSIONS

Nowadays, many food processing technologies are used in industries that meet consumer requirements, ensure food safety, and maintain nutritional value. These techniques differ in terms of heat, time, and energy required for processing. Thermal processing technologies are standard technologies to improve the shelf-life of dairy products. However, these technologies are related to irreversible changes in the treated products' sensory and nutritive profile. On the other hand, non-thermal processing techniques have been proven to produce food while preserving nutritional quality and value without a chemical additive. The non-thermal techniques aim to inhibit microorganisms and enzymes from reducing the shelf-life of dairy products without using heat treatments. These techniques are environmentally friendly due to reducing energy and water consumption and gas emissions. Using renewable energy resources such as hydroelectric power could save energy and processing cost.

REFERENCES

Aadil, R. M., Zeng, X.-A., Wang, M.-S., Liu, Z.-W., Han, Z., Zhang, Z.-H., Hong, J., & Jabbar, S. (2015). A potential of ultrasound on minerals, micro-organisms, phenolic compounds and colouring pigments of grapefruit juice. *International Journal of Food Science and Technology*, 50(5), 1144–1150. https://doi.org/10.1111/ijfs.12767

Administration, U. S. E. I. (2006). *Annual Energy Review 2005*. Energy Information Administration.

Ajmal, M., Nadeem, M., Imran, M., & Junaid, M. (2018). Lipid compositional changes and oxidation status of ultra-high temperature treated Milk. *Lipids in Health and Disease*, 17(1), 227. https://doi.org/10.1186/s12944-018-0869-3

Arya, S. S., Sawant, O., Sonawane, S. K., Show, P. L., Waghamare, A., Hilares, R., & Santos, J. C. Dos (2020). Novel, nonthermal, energy efficient, industrially scalable hydrodynamic cavitation—Applications in food processing. *Food Reviews International*, 36(7), 668–691. https://doi.org/10.1080/87559129.2019.1669163

Awasti, N., Chaudhary, P., & Anand, S. (2019). Manufacturing low-spore-count skim milk powders by combining optimized raw milk holding conditions and hydrodynamic cavitation. *American Dairy Science Association Annual Meeting*, 9–10.

Awasti, N., Sunkesula, V., & Hammam, A. (2021). Current and emerging food safety systems and tools to prevent food hazards in the food processing industry. In: *Biological and Chemical Hazards in Food and Food Products*. Apple Academic Press (In press).

Barnett, S. (1998). Nonthermal issues: Cavitation—Its nature, detection and measurement. *Ultrasound in Medicine and Biology*, 24, S11–S21. https://doi.org/10.1016/S0301-5629(98)00074-X

Barrett, D. M., Beaulieu, J. C., & Shewfelt, R. (2010). Color, flavor, texture, and nutritional quality of fresh-cut fruits and vegetables: Desirable levels, instrumental and sensory measurement, and the effects of processing. *Critical Reviews in Food Science and Nutrition*, 50(5), 369–389. https://doi.org/10.1080/10408391003626322

Batie, S. S., & Healy, R. G. (1983). The future of American agriculture. *Scientific American*, 248(2), 45–53.

Brush, A., Masanet, E., Worrell, E. (2011). Energy efficiency improvement and cost saving opportunities for dairy processing industry – An Energy Star guide for energy and plant managers. Lawrence Berkeley National Laboratory, Environmental Energy Technologies Division, Energy Analysis Department (online report), Berkeley, CA, USA. http://www.energystar.gov/index.cfm?c=in_focus.bus_dairy_processing.

Cacace, F., Bottani, E., Rizzi, A., & Vignali, G. (2020). Evaluation of the economic and environmental sustainability of high pressure processing of foods. *Innovative Food Science and Emerging Technologies*, 60. https://doi.org/10.1016/j.ifset.2019.102281. 102281

Caminiti, I. M., Noci, F., Muñoz, A., Whyte, P., Morgan, D. J., Cronin, D. A., & Lyng, J. G. (2011). Impact of selected combinations of non-thermal processing technologies on the quality of an apple and cranberry juice blend. *Food Chemistry*, 124(4), 1387–1392. https://doi.org/10.1016/j.foodchem.2010.07.096

Chambers, L. A., & Gaines, N. (1932). Some effects of intense audible sound on living organisms and cells. *Journal of Cellular and Comparative Physiology, 1*(3), 451–473. https://doi.org/10.1002/jcp.1030010310

Chaudhary, P. (2019). Hydrodynamic Cavitation as In-line Process to Control Common Dairy Sporeformers. *Electronic Theses and Dissertations.* 3370. https://openprairie.sdstate.edu/etd/3370.

Cheftel, J. C. (1992). Effects of high hydrostatic pressure on food constituents: An overview. *Hiph Pressure Biotechnology,* 195–209.

Cheigh, C.-I., Park, M.-H., Chung, M.-S., Shin, J.-K., & Park, Y.-S. (2012). Comparison of intense pulsed light- and ultraviolet (UVC)-induced cell damage in Listeria monocytogenes and Escherichia coli O157:H7. *Food Control, 25*(2), 654–659. https://doi.org/10.1016/j.foodcont.2011.11.032

Chemat, F., Huma, Z., & Khan, M. K. (2011). Applications of ultrasound in food technology: Processing, preservation and extraction. *Ultrasonics Sonochemistry, 18*(4), 813–835. https://doi.org/10.1016/j.ultsonch.2010.11.023

Clark, J. P. (2006). Processing-pulsed electric field processing. *Food Technology-Chicago, 60*(1), 66–67.

Compton, M., Willis, S., Rezaie, B., & Humes, K. (2018). Food processing industry energy and water consumption in the Pacific northwest. *Innovative Food Science and Emerging Technologies, 47,* 371–383. https://doi.org/10.1016/j.ifset.2018.04.001

Cullen, P. J., Valdramidis, V. P., Tiwari, B. K., Patil, S., Bourke, P., & O'Donnell, C. P. (2010). Ozone processing for food preservation: An overview on fruit juice treatments. *Ozone: Science and Engineering, 32*(3), 166–179. https://doi.org/10.1080/01919511003785361

Dalsgaard, H., & Abbotts, W. (2003). Improving energy efficiency. In: *Environmentally-Friendly Food Processing* (pp. 116–129). Elsevier.

De Stoutz, W. P. (1976). Method and apparatus for the irradiative treatment of beverages. Google Patents (Patent No. US3934042A)

Einstein, D., Worrell, E., & Khrushch, M. (2001). *Steam Systems in Industry: Energy Use and Energy Efficiency Improvement Potentials.* Lawrence Berkeley National Laboratory. Available at https://escholarship.org/uc/item/3m1781f1.

Farkas, J. (2006). Irradiation for better foods. *Trends in Food Science and Technology, 17*(4), 148–152. https://doi.org/10.1016/j.tifs.2005.12.003

Feliciano, C. P., De Guzman, Z. M., Tolentino, L. M. M., Cobar, M. L. C., & Abrera, G. B. (2014). Radiation-treated ready-to-eat (RTE) chicken breast Adobo for immuno-compromised patients. *Food Chemistry, 163,* 142–146. https://doi.org/10.1016/j.foodchem.2014.04.087

Fernandes, Â., Antonio, A. L., Oliveira, M. B. P. P., Martins, A., & Ferreira, I. C. F. R. (2012). Effect of gamma and electron beam irradiation on the physico-chemical and nutritional properties of mushrooms: A review. *Food Chemistry, 135*(2), 641–650. https://doi.org/10.1016/j.foodchem.2012.04.136

Garcia-Gonzalez, L., Geeraerd, A. H., Spilimbergo, S., Elst, K., Van Ginneken, L., Debevere, J., Van Impe, J. F., & Devlieghere, F. (2007). High pressure carbon dioxide inactivation of microorganisms in foods: The past, the present and the future. *International Journal of Food Microbiology, 117*(1), 1–28. https://doi.org/10.1016/j.ijfoodmicro.2007.02.018

Gayán, E., García-Gonzalo, D., Álvarez, I., & Condón, S. (2014). Resistance of Staphylococcus aureus to UV-C light and combined UV-heat treatments at mild temperatures. *International Journal of Food Microbiology, 172,* 30–39. https://doi.org/10.1016/j.ijfoodmicro.2013.12.003

Gillham, C. R., Fryer, P. J., Hasting, A. P. M., & Wilson, D. I. (2000). Enhanced cleaning of whey protein soils using pulsed flows. *Journal of Food Engineering, 46*(3), 199–209

Glasson, J., & Therivel, R. (2019). *Introduction to Environmental Impact Assessment,* 5th Edition. Routledge, London. https://doi.org/10.4324/9780429470738.

Guerrero-Beltran, J. A., & Barbosa-Canovas, G. V. (2005). Reduction of Saccharomyces cerevisiae, Escherichia coli and Listeria innocua in apple juice by ultraviolet light. *Journal of Food Process Engineering, 28*(5), 437–452. https://doi.org/10.1111/j.1745-4530.2005.00040.x

Han, J. (2009). Food safety and innovative food packaging. In: *Microbiologically Safe Foods* John, H. (Ed.) (pp. 507–521). John Wiley & Sons, Inc. https://doi.org/10.1002/9780470439074.ch25

Hansen, L. G. (1999). Environmental regulation through voluntary agreements. In: *Voluntary Approaches in Environmental Policy* (pp. 27–54). Springer, Netherlands. https://doi.org/10.1007/978-94-015-9311-3_3.

Huang, H.-W., Wu, S.-J., Lu, J.-K., Shyu, Y.-T., & Wang, C.-Y. (2017). Current status and future trends of high-pressure processing in food industry. *Food Control, 72*, 1–8. https://doi.org/10.1016/j.foodcont.2016.07.019

Katariya, P., Arya, S. S., & Pandit, A. B. (2020). Novel, non-thermal hydrodynamic cavitation of orange juice: Effects on physical properties and stability of bioactive compounds. *Innovative Food Science and Emerging Technologies, 62*. https://doi.org/10.1016/j.ifset.2020.102364. 102364

Kebede, B. T., Grauwet, T., Palmers, S., Vervoort, L., Carle, R., Hendrickx, M., & Van Loey, A. (2014). Effect of high pressure high temperature processing on the volatile fraction of differently coloured carrots. *Food Chemistry, 153*, 340–352. https://doi.org/10.1016/j.foodchem.2013.12.061

Knorr, D., Froehling, A., Jaeger, H., Reineke, K., Schlueter, O., & Schoessler, K. (2011). Emerging technologies in food processing. *Annual Review of Food Science and Technology, 2*(1), 203–235. https://doi.org/10.1146/annurev.food.102308.124129

Kumari, A., & Farid, M. (2020). Optimization of high pressure processing for microbial load reduction in Diospyros kaki 'Fuyu' pulp using response surface methodology. *Journal of Food Science and Technology, 57*(7), 2472–2479. https://doi.org/10.1007/s13197-020-04282-z

Lammerding, A. M., & Doyle, M. P. (1990). Stability of Listeria monocytogenes to non-thermal processing conditions. In: *Foodborne Listeriosis: Topics in Industrial Microbiology. Volume 2* A. J. Miller, J. L. Smith, & G. A. Somkuti (Eds.). Elsevier Science Publishers, Journals Division, New York, NY.

Larkin, J. W., & Spinak, S. H. (1996). Safety considerations for ohmically heated, aseptically processed, multiphase low-acid food products: Ohmic heating for thermal processing of foods: Government, industry, and academic perspectives. *Food Technology (Chicago), 50*(5), 242–245

Leistner, L., & Gould, G. W. (2002). Hurdle technologies. In: *Hurdle Technologies*. Springer, US. https://doi.org/10.1007/978-1-4615-0743-7.

Li, K., Woo, M. W., Patel, H., Metzger, L., & Selomulya, C. (2018). Improvement of rheological and functional properties of milk protein concentrate by hydrodynamic cavitation. *Journal of Food Engineering, 221*, 106–113. https://doi.org/10.1016/j.jfoodeng.2017.10.005

Lima, M., Zhong, T., & Lakkakula, N. R. (2002). Ohmic heating: A value-added food processing tool. *Louisiana Agriculture, 45*(4), 16.

Lung, R. B., Masanet, E., & McKane, A. (2006). *The Role of Emerging Technologies in Improving Energy Efficiency: Examples from the Food Processing Industry*. https://escholarship.org/uc/item/43c841xs

Martín-Belloso, O., Soliva-Fortuny, R., Elez-Martínez, P., Robert Marsellés-Fontanet, A., & Vega-Mercado, H. (2014). Non-thermal processing technologies. In: *Food Safety Management* Y. Motarjemi & H. B. T.-F. S. M. editors Lelieveld (Eds.). (pp. 443–465). Elsevier. https://doi.org/10.1016/B978-0-12-381504-0.00018-4.

Masanet, E. (2008). *Energy Efficiency Improvement and Cost Saving Opportunities for the Fruit and Vegetable Processing Industry. An Energy Star Guide for Energy and Plant Managers*. https://escholarship.org/uc/item/8h25n5pr

Masanet, E., Brush, A., & Worrell, E. (2014). Energy efficiency opportunities in the US Dairy processing industry. *Energy Engineering, 111*(5), 7–34.

Merrill, R. A. (1997). Food safety regulation: Reforming the Delaney clause. *Annual Review of Public Health, 18*(1), 313–340.

Mir, S. A., Shah, M. A., & Mir, M. M. (2016). Understanding the role of plasma technology in food industry. *Food and Bioprocess Technology, 9*(5), 734–750.

Moraru, C. I., & Schrader, E. U. (2008). Applications of membrane separation in the brewing industry. In: A. K. Pabby, S. S. H. Rizvi, & Requena, A. M. S. (Eds.) *Handbook of Membrane Separations* (pp. 557–583). CRC Press, Boca Raton, FL.

Muller, D. C. A., Marechal, F. M. A., Wolewinski, T., & Roux, P. J. (2007). An energy management method for the food industry. *Applied Thermal Engineering, 27*(16), 2677–2686.

Oupadissakoon, G., Chambers, D. H., & Chambers, E. (2009). Comparison of the sensory properties of ultra-high-temperature (UHT) milk from different countries. *Journal of Sensory Studies, 24*(3), 427–440. https://doi.org/10.1111/j.1745-459X.2009.00219.x

Pereira, R. N., & Vicente, A. A. (2010). Environmental impact of novel thermal and non-thermal technologies in food processing. *Food Research International, 43*(7), 1936–1943.

Qin, B.-L., Chang, F.-J., Barbosa-Cánovas, G. V., & Swanson, B. G. (1995). Nonthermal inactivation of Saccharomyces cerevisiae in apple juice using pulsed electric fields. *LWT—Food Science and Technology, 28*(6), 564–568. https://doi.org/10.1016/0023-6438(95)90002-0

Qin, B. L., Pothakamury, U. R., Barbosa-Cánovas, G. V., & Swanson, B. G. (1996). Nonthermal pasteurization of liquid foods using high-intensity pulsed electric fields. *Critical Reviews in Food Science and Nutrition, 36*(6), 603–627. https://doi.org/10.1080/10408399609527741

Raso, J., & Barbosa-Cánovas, G. V. (2003). Nonthermal preservation of foods using combined processing techniques. *Critical Reviews in Food Science and Nutrition, 43*(3), 265–285. https://doi.org/10.1080/10408690390826527

Reid, K. C., & Jackson, P. (1956). Non-thermal drying of brown marine algae. *Journal of the Science of Food and Agriculture, 7*(4), 291–300. https://doi.org/10.1002/jsfa.2740070413

Rendueles, E., Omer, M. K., Alvseike, O., Alonso-Calleja, C., Capita, R., & Prieto, M. (2011). Microbiological food safety assessment of high hydrostatic pressure processing: A review. *LWT—Food Science and Technology, 44*(5), 1251–1260. https://doi.org/10.1016/j.lwt.2010.11.001

Rentschler, H. C., Nagy, R., & Mouromseff, G. (1941). Bactericidal effect of ultraviolet radiation. *Journal of Bacteriology, 41*(6), 745–774

Salve, A. R., Pegu, K., & Arya, S. S. (2019). Comparative assessment of high-intensity ultrasound and hydrodynamic cavitation processing on physico-chemical properties and microbial inactivation of peanut milk. *Ultrasonics Sonochemistry, 59*, 104728. https://doi.org/10.1016/j.ultsonch.2019.104728

Sampedro, F., McAloon, A., Yee, W., Fan, X., & Geveke, D. J. (2014). Cost analysis and environmental impact of pulsed electric fields and high pressure processing in comparison with thermal pasteurization. *Food and Bioprocess Technology, 7*(7), 1928–1937. https://doi.org/10.1007/s11947-014-1298-6

Sandu, C., & Singh, R. K. (1991). Energy increase in operation and cleaning due to heat-exchanger fouling in milk pasteurization. *Food Technology (USA), 45*(12), 84–91.

Shao, Y., Zhu, S., Ramaswamy, H., & Marcotte, M. (2010). Compression heating and temperature control for high-pressure destruction of bacterial spores: An experimental method for kinetics evaluation. *Food and Bioprocess Technology, 3*(1), 71.

Sharma, P., Bremer, P., Oey, I., & Everett, D. W. (2014). Bacterial inactivation in whole milk using pulsed electric field processing. *International Dairy Journal, 35*(1), 49–56. https://doi.org/10.1016/j.idairyj.2013.10.005

Sluder, J. C., Olsen, R. W., & Kenyon, E. M. (1947). A method for the production of dry powdered orange juice. *Food Technology, 1*(1), 85–94.

Tani, A., Ono, Y., Fukui, S., Ikawa, S., & Kitano, K. (2012). Free radicals induced in aqueous solution by non-contact atmospheric-pressure cold plasma. *Applied Physics Letters, 100*(25). https://doi.org/10.1063/1.4729889. 254103

Thirumdas, R., Sarangapani, C., & Annapure, U. S. (2015). Cold plasma: A novel Non-Thermal Technology for Food processing. *Food Biophysics, 10*(1), 1–11. https://doi.org/10.1007/s11483-014-9382-z

Toepfl, S., Mathys, A., Heinz, V., & Knorr, D. (2006). Review: Potential of high hydrostatic pressure and pulsed electric fields for energy efficient and environmentally friendly food processing. *Food Reviews International, 22*(4), 405–423.

Toogood, S., & Key, M. (2000). *Air pollution. Food Industry and the Environment in the European Union, Practical Issues and Cost Implications* (2nd ed.). Aspen Publishers Inc, Gaithersburg, MD.

Toyoda, I., Schreier, P. J. R., & Fryer, P. J. (1994). A computational model for reaction fouling from whey protein solutions. In: *Proceedings of Fouling Cleaning in Food Processing* (pp. 222–229).

Unluturk, S., Atılgan, M. R., Handan Baysal, A., & Tarı, C. (2008). Use of UV-C radiation as a non-thermal process for liquid egg products (LEP). *Journal of Food Engineering, 85*(4), 561–568. https://doi.org/10.1016/j.jfoodeng.2007.08.017

U.S. Census Bureau. 2009. *Statistical Abstract of the United States, 2000*. Washington, DC: Government Printing Office.

US Department of Commerce, B. of the C. (1980). *Annual Survey of Manufactures: Statistics for Industry Groups and Industries. Volume M80 (AS)-1*. US Government Printing Office.

Vicente, A. A., Castro, I. de, & Teixeira, J. A. (2006). *Ohmic Heating for Food Processing.* http://hdl.handle.net/1822/48461

Wang, L. (2008). *Energy Efficiency and Management in Food Processing Facilities.* CRC Press, Boca Raton, FL.

Wertheim, J. H. (1960). Radiation sterilization of fluid food products. Google Patents (Patent No. US2962380A).

Wu, D., Forghani, F., Daliri, E. B.-M., Li, J., Liao, X., Liu, D., Ye, X., Chen, S., & Ding, T. (2020). Microbial response to some nonthermal physical technologies. *Trends in Food Science and Technology, 95,* 107–117. https://doi.org/10.1016/j.tifs.2019.11.012

Zhang, Z.-H., Wang, L.-H., Zeng, X.-A., Han, Z., & Brennan, C. S. (2019). Non-thermal technologies and its current and future application in the food industry: A review. *International Journal of Food Science and Technology, 54*(1), 1–13. https://doi.org/10.1111/ijfs.13903

Zhao, Y.-M., de Alba, M., Sun, D.-W., & Tiwari, B. (2019). Principles and recent applications of novel non-thermal processing technologies for the fish industry—A review. *Critical Reviews in Food Science and Nutrition, 59*(5), 728–742. https://doi.org/10.1080/10408398.2018.1495613

Index

A

Anhydrous milk fat, 100
Applications of UV light in dairy, 75
    disinfection of air, 75
    disinfection of water, 75
    surface applications, 75

B

Bacteriocin application, 116
Bacteriocin, 115
Beta-serum and WPPC, 101
Biopreservation, 115
Butter, 97
Butter milk derivatives, 101
Butter oil, 98

C

Cheese, 39, 59, 98
Chemical preservatives, 112
Cold plasma, 4, 44, 143
    dairy based beverages, 61
    effect, 147
    food component modification, 45
    mechanism, 145
    microbial load inactivation, 45
    milk, 47
    milk protein, 60
    scope, 151
Consumer perception, 107

E

Efficacy of UV, 75
Electrochemical methods, 132
    electrocoagulation, 135
    electroflotation, 132
Emulsification, 84
Energy savings, 166
Enterocins, 121
Environmental impact, 165

F

Fermentation, 89
Fluid milk, 95
Foaming, 88
Food safety, 107

G

Gelation, 88

H

High hydrostatic pressure processing, 35
Homogenization, 85
Hydrodynamic cavitation, 132

I

Ionizing radiation sources, 106
Irradiation effects, 108
Irradiation technique in dairy applications, 108

L

Lacticins, 120
Low temperature high pressure processing, 4

M

Market milk, 36
Membrane processing, 131
Milk shake, 36
    high-pressure processing on antioxidative capacity, 37
    high-pressure processing on total phenol content, 37

N

Nisin, 116

P

Pediocins, 119
PEF processing of milk, 22
    PEF treatment on milk enzymes, 24
    PEF treatment on milk lipids, 26
    PEF treatment on milk proteins, 28
    PEF treatment on vitamins, 25
Pulsed electric field, 14
    equipment, 18
    introduction, 14
    mechanisms of microbial inactivation, 16
    PEF in combination with other technologies, 21
    principle, 15
Pulsed light, 5

R

Regulatory requirements, 162

S

Sonocrystalization, 87
Sources of uv light technology, 70
    excimer lamps, 71
    light-emitting diodes, 72
    mercury lamps, 71
    pulsed UV lamps, 71
    UV light devices, 72
Supercritical CO_2 extraction, 94
Supercritical process, 95

U

Ultrasonication, 3
Ultrasound, 81
    applications, 84
    principles, 82
Ultraviolet light, 1, 67

W

Waste water from dairy, 128
    characteristics, 128
    sources, 128
    treatment, 129

Y

Yogurt, 38

Printed in the United States
by Baker & Taylor Publisher Services